THE EVERYTHING
GUIDE TO
ALGEBRA

Dear Reader,

I first heard the joke "Q: Define aftermath. A: Algebra." very early in my teaching career, when the only people who studied elementary algebra were the "college-bound" students, and the only ones likely to take intermediate algebra were those who were headed toward physics, engineering, or mathematics.

The times have changed. Since the introduction of the scientific calculator, the personal computer, and the graphing calculator, the changes in educational expectations have become much more intense. The PC did not quite live up to its claim that users would have more free time. Instead, the amount of data that was available to the general public grew exponentially. Employers expected that more of their employees would access and analyze the data. With this increase in expectations came a significant increase in the number of students taking more than elementary algebra.

You will find that this book uses the graphing calculator with some regularity. This text also contains traditional symbol manipulation, because these skills are the building blocks for more advanced topics.

I hope that this book helps you better understand some of these basic principles of algebra and that you enjoy working through the material.

Chris Monahan

Welcome to the EVERYTHING® Series!

These handy, accessible books give you all you need to tackle a difficult project, gain a new hobby, comprehend a fascinating topic, prepare for an exam, or even brush up on something you learned back in school but have since forgotten.

You can choose to read an *Everything*® book from cover to cover or just pick out the information you want from our four useful boxes: e-rules, e-questions, e-alerts, and e-ssentials.

We give you everything you need to know on the subject, but throw in a lot of fun stuff along the way, too.

We now have more than 400 *Everything*® books in print, spanning such wide-ranging categories as weddings, pregnancy, cooking, music instruction, foreign language, crafts, pets, New Age, and so much more. When you're done reading them all, you can finally say you know *Everything*®!

RULE

Important rules
to remember

QUESTION

Answers to
common questions

ALERT

Urgent
warnings

ESSENTIAL

Quick
handy tips

PUBLISHER Karen Cooper

DIRECTOR OF ACQUISITIONS AND INNOVATION Paula Munier

MANAGING EDITOR, EVERYTHING® SERIES Lisa Laing

COPY CHIEF Casey Ebert

ASSISTANT PRODUCTION EDITOR Jacob Erickson

ACQUISITIONS EDITOR Lisa Laing

SENIOR DEVELOPMENT EDITOR Brett Palana-Shanahan

EDITORIAL ASSISTANT Ross Weisman

EVERYTHING® SERIES COVER DESIGNER Erin Alexander

LAYOUT DESIGNER Erin Dawson

Visit the entire Everything® series at *www.everything.com*

THE
EVERYTHING®
GUIDE TO ALGEBRA

A step-by-step guide to the basics
of algebra—*in plain English!*

Christopher Monahan

Former President, Association of Mathematics Teachers of New York State

adamsmedia
Avon, Massachusetts

To my family
dht, kem-s, rmm-s, kmm, art, bht, sast, and crt

An Everything® Series Book.
Everything® and everything.com® are registered trademarks of F+W Media, Inc.

Published by Adams Media, a division of F+W Media, Inc.
57 Littlefield Street, Avon, MA 02322 U.S.A.
www.adamsmedia.com

ISBN 10: 1-4405-0458-X
ISBN 13: 978-1-4405-0458-7
eISBN 10: 1-4405-0459-8
eISBN 13: 978-1-4405-0459-4

Printed in the United States of America.

10 9 8 7 6 5 4 3 2 1

Library of Congress Cataloging-in-Publication Data
is available from the publisher.

This publication is designed to provide accurate and authoritative information with regard to the subject matter covered. It is sold with the understanding that the publisher is not engaged in rendering legal, accounting, or other professional advice. If legal advice or other expert assistance is required, the services of a competent professional person should be sought.

—From a *Declaration of Principles* jointly adopted by a Committee of the American Bar Association and a Committee of Publishers and Associations

Many of the designations used by manufacturers and sellers to distinguish their products are claimed as trademarks. Where those designations appear in this book and Adams Media was aware of a trademark claim, the designations have been printed with initial capital letters.

This book is available at quantity discounts for bulk purchases.
For information, please call 1-800-289-0963.

Contents

Acknowledgments

The material and examples used in this book are drawn from the interaction with the thousands of students who sat in my classes every day for more than thirty years. I thank you for all you taught me about teaching, and I hope this book makes you proud.

Thanks go to Dave Bock, who put me in contact with my agent, Grace Freedson. I also must thank Lisa Laing for her expertise in editing this text and Matt LeBlanc for his advice on the graphics. I especially want to thank my wife Diane. Without her support and encouragement, this book would not have been written.

Top 10 Reasons Why—And How—You Should Study Algebra

1. Mathematicians are the original texters. There really is no difference between using statements like LOL and BFF to save space and using x for the number that is unknown. Don't let the use of symbols scare you.

2. Mathematics is not as rigid as people think. There are many ways to solve problems. Algebraic, graphical, and tabular approaches are often equally good options for getting to the answer.

3. Algebra has many applications outside of the mathematics classroom.

4. You use mathematics more than you think you do—you just call it reasoning. Don't be afraid to apply common sense and your own experiences to solving problems.

5. Using technology in studying mathematics is here to stay. More problems can be solved through the application of graphing calculators and computer programs.

6. Most fields of study have a mathematical component to them. Very often, this component is statistics. Algebra is a basic tool in the study of statistics.

7. Remember that x is only one letter in the alphabet. When choosing a variable, use a letter that is related to the value sought (n for number, t for time, v for volume, and r for radius are all good choices).

8. Some of the best learning happens right after you make a mistake. Work through the exercises in the text before working on the exercises at the end of the chapter. Make sure that you understand the process being discussed.

9. This is a pencil-and-paper book. Make sure that you have plenty of paper and a writing implement with you as you read. Do the exercises in the section before you look at the exercises at the end of the chapter.

10. You will enjoy your success when you have finished the text and have more confidence in your ability to use algebra.

Introduction

IT IS FAIRLY COMMON knowledge that geometry has its roots in Egypt and Greece. The Egyptians used geometry for practical purposes, and the Greeks formalized the study of geometry (and of most other subjects) through their use of logic. What we know as the Pythagorean Theorem, perhaps the most famous of all the formulas in geometry, was actually being used in Egypt well before Pythagoras was born and was being used in China centuries before Pythagoras. Much of what is studied in mathematics has a European slant to it because that is where the founders of the Western world originated. But algebra was developed and refined in India and along the sub-Asian continent as a problem-solving tool while Europe was living through the Dark Ages. Foremost in its development was the mathematician al-Kwarizimi, who lived 1,200 years ago in Baghdad.

During the European Renaissance and in the years that followed, mathematicians such as Cardano, Tartaglia, Fibonacci, Descartes, Leibnitz, Newton, Pascal, Euler, and the Bernoullis used algebra to solve more interesting and complex problems. Geometry, calculus, probability, and statistics are topics studied in today's high school mathematics classes. Each branch was developed to yield more information about the world—as it is, and as it might be. This may be the most important piece of information for you to keep in mind as you work through (note that we say, "work through," not just "read") this book. Mathematics has, for the most part, been developed to solve problems that have not been solved. Sometimes, the motivation is pride (as in the cases of Cardano and Tartaglia); sometimes it is financial, as in the case of Fibonacci. But very often physics is a major motivator.

The human drive to be more efficient in the workplace requires a better understanding of the forces that machines and nature exert. Leibnitz and Newton developed calculus on the basis of the analytic geometry created by Descartes, and their work was improved upon by Euler, the Bernoullis, and a host of other mathematicians, including many who are still continuing their investigations today. Pascal, who designed the first mechanical calculator (which was too complicated to be engineered with the technology of his day), is better remembered for formalizing the laws of probability. All of the work done by these men had its basis in algebra.

Just as the basic addition facts are the fundamental building blocks for elementary and middle school mathematics, algebra is the basic tool of high school mathematics, and calculus serves that purpose for higher-level mathematics. This book will serve as an extra resource for those who are currently studying elementary and intermediate algebra. It can also offer a review for those who need to refresh their skills.

Technology is a tool often applied in the study of algebra, and it is used in this text when appropriate. Appendix A gives a general overview of the topic, but the nuts and bolts of deriving the equations of best fit are included in the chapters themselves. The two tools used in writing this text are the TI 84 and TI Nspire calculators. An indispensable application of technology is its use to model behavior from a set of data with an equation of best fit. This, too, is included in the text, and it may be a new area of study for those using this book for review.

CHAPTER 1

The Basic Building Blocks

In math, as in any form of communication, there are rules that are agreed upon so that everyone can understand exactly what is being communicated. Oral and written languages use vocabulary, grammar, and sentence structure to communicate effectively. Mathematics uses its own form of these entities as well. Learning the language of mathematics is a key aspect of understanding the concepts being communicated. Numbers (analogous in some ways to the alphabet in which a language is written) and how these numbers are combined (much as letters are combined in the spelling of words) are the foundation for the study of algebra. This chapter will introduce the basic building blocks of algebra and give you a sturdy foundation for your future study.

Real Numbers and Subsets

Children learn about numbers quickly. Youngsters are taught to hold up their fingers when asked their age. They learn quickly to notice whether they received the same number of cookies as their siblings during snack time. Counting comes naturally. The **natural numbers** (or counting numbers) are 1, 2, 3, 4, …. These are the basic values from which we start.

Children learn about zero when they do not get what they asked for. Including zero with the set of natural numbers gives you the set of **whole numbers**. Students, especially those from colder climates, learn about the negative numbers when "it is too cold to play outside."

When combined, the whole numbers and their negatives make up the set of **integers**. Children encounter fractions when they break their cookies apart or share their toys with a playmate. The formal name for fractions, **rational numbers**, comes from the idea that a fraction is the **ratio** of two integers. You can see that children have been exposed to a fair amount of mathematics before they have even entered their first classroom!

ESSENTIAL

Having learned about square roots of the "nice" numbers (1, 4, 9, 16, 25, and so on), students are often confused when their teacher asks about the square root of 3. There is some relief in the class when they learn that numbers such as the square root of 3 are **irrational**. For the record, the numbers are irrational merely because they are not rational. That is, irrational numbers cannot be written as the ratio of two integers.

There is a hierarchy for these sets of numbers. All of the natural numbers are included in the set of whole numbers; all of the whole numbers are included in the set of integers; and all of the integers are included in the set of rational numbers. Of course, there are no numbers that are common to both the set of rational numbers and the set of irrational numbers. However, when these two sets are combined (this is called taking the **union** of the sets), a new set of numbers is formed, the set of **real numbers**. A simple explanation for the set of real numbers is that it represents all the numbers that can be placed on a number line. A graphical representation for the relationship among these sets of numbers follows.

**Set of Real Numbers
and Its Subsets**

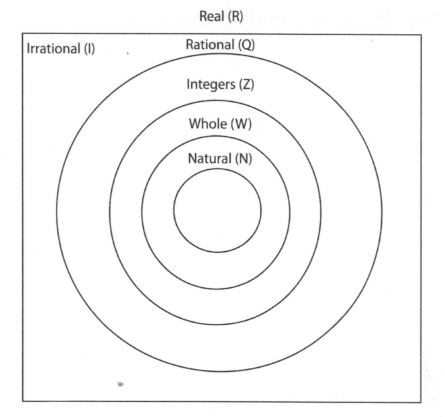

The number 8, for example, is a counting number (N), a whole number (W), an integer (Z), a rational number (Q), and a real number (R). $\frac{3}{5}$ is a rational number (Q) and a real number (R), but it belongs to none of the other categories. $\sqrt{3}$ is an irrational number (I) and a real number (R).

Integer Arithmetic

A geometric approach to the arithmetic of integers will tie together what you have known for a long time, will reinforce or clarify what may have been a fuzzy rule, and will, we hope, make sense. Start with a simple example: $2+3=5$. You can recall any number of different manipulatives (chips, coins, marbles, or blocks, for instance) that helped you with this problem when you were younger. The following figure illustrates another way in which this problem may have been shown to you.

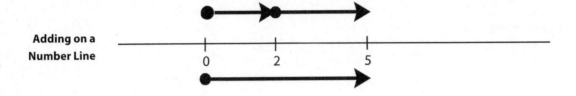

Adding on a Number Line

Note that the starting point of the second number lines up with the arrow tip of the first number. Accordingly, the next figure shows the sum 2 + (-3).

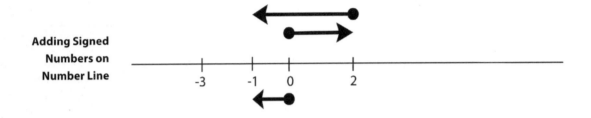

Adding Signed Numbers on Number Line

To show subtraction, you need to realize that the direction for the second arrow must be reversed. The diagram for $5 - 2$ is shown in the following figure, as is the sum $5 + (-2)$. Yes, the subtraction of a positive number is the same as the addition of the first number and the negative of the second number.

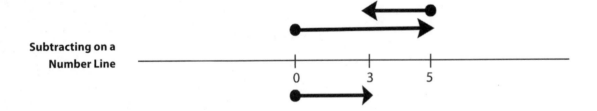

Subtracting on a Number Line

Because subtraction requires that you change the direction of the arrow, and because the arrow for a negative number points to the left, $5 - (-2)$ looks the same as $5 + 2$.

Recall that to multiply $2 * 3$, you were taught to think of taking the number 2 and adding it 3 times. The following figure shows this.

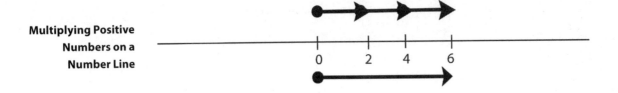

Multiplying Positive Numbers on a Number Line

In the same manner, -2 ∗ 3 requires that we take -2 and add it 3 times, as shown in the next figure.

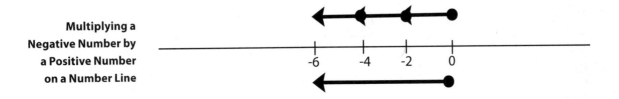

Multiplying a Negative Number by a Positive Number on a Number Line

Because multiplication is known to be commutative (2 ∗ 3 = 3 ∗ 2), it should also be the case that 3 ∗ (-2) is the same as -6. But how do you interpret the geometry? The length 3 is written twice to get 6, but the direction of the product must be rotated to the other side of the number line to account for the negative result, as shown in this figure.

Multiplying a Negative Number by a Positive Number on a Number Line

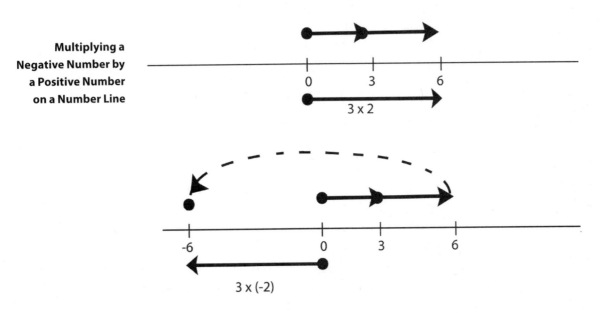

The last rule to examine is the product of two negative numbers, such as (-2) * (-3). The directed segment -2 must be written three times, and the result is rotated to the other side of the number line to get 6, as shown in the following figure.

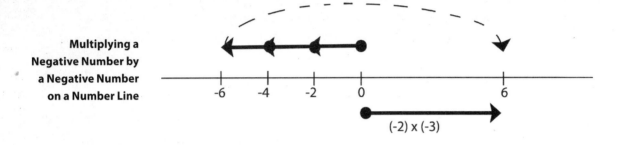

Multiplying a Negative Number by a Negative Number on a Number Line

(-2) x (-3)

Bases and Exponents

What is the value of $3 \times 3 \times 3 \times 3$? One possible response you may have when you read this problem is "3×3 is 9, 9×3 is 27, and 27×3 is 81. The answer is 81." Mathematicians have developed a shorthand notation for writing the product of the same number multiple times: ***exponents***. In this example, the number 3 is used as a *factor* four times. The short hand notation for this problem is $3 \times 3 \times 3 \times 3 = 3^4$. We say that 3 is the *base* and 4 is the exponent. What is the value of 2^5? Well, 2^5 means to use 2 as a factor 5 times, so the product is 32. The first definition for an exponent is that, as a positive integer, the exponent indicates the number of times the base should be used as a factor.

$$8 \times 8 \times 8 \times 8 \times 8 \times 8 \times 8 \times 8 \times 8 \times 8 = 8^{10} = 1,073,741,824$$

$$\frac{2}{3} \times \frac{2}{3} \times \frac{2}{3} \times \frac{2}{3} \times \frac{2}{3} \times \frac{2}{3} \times \frac{2}{3} = \left(\frac{2}{7}\right)^7 = \frac{128}{2187}$$

$$(5)(5)(5)(5)(5)(5) = 5^6 = 15,625$$

Please note that you can write multiplication problems using a set of parentheses, rather than using $\times$ or $*$ to represent the operation.

Order of Operations

If the set of real numbers is the alphabet for elementary algebra, then the order of operations are the first rules used for putting these blocks into a meaningful order. The order in which we carry out the operations that connect our numbers is as follows:

- **P**arentheses (or any other grouping process, such as brackets or braces)
- **E**xponents
- **M**ultiplication and **D**ivision, as they occur from left to right in the problem
- **A**ddition and **S**ubtraction, as they occur from left to right

PEMDAS – **P**lease **E**xcuse **M**y **D**ear **A**unt **S**ally is mnemonic device for remembering the order of operations.

Calculate the value: $8 + 9(12 + 5) - 8^2 + 85 \div 17$

Parentheses	$8 + 9(\mathbf{17}) - 8^2 + 85 \div 17$
Exponents	$8 + 9(17) - \mathbf{64} + 85 \div 17$
Multiplication and **D**ivision	$8 + \mathbf{153} - 64 + \mathbf{5}$
Addition and **S**ubtraction	$\mathbf{161} - 64 + 5$
	$\mathbf{97} + 5$
	102

Calculate the value: $120 - 5(8 + 12^2 - 2 \times 9) \div 67$

$$120 - 5(8 + 144 - 2 \times 9) \div 67$$
$$120 - 5(8 + 144 - 18) \div 67$$
$$120 - 5(134) \div 67$$
$$120 - 670 \div 67$$
$$120 - 10$$
$$110$$

A more difficult example is $\left(\dfrac{3^3+5^2}{40-6^2}\right)^2$ because it is hard to read the problem "from left to right." The translation of this problem into one that is clearer to read gives you

$$((3^3+5^2)\div(40-6^2))^2 = ((27+25)\div(40-36))^2$$
$$=(52\div4)^2$$
$$=(13)^2=169$$

Properties of Operations

The human brain, which is capable of incredibly complex thought processes, is a very simple organ when it comes to computation. Like a computer, the brain can deal with only two numbers at a time. The speed with which some people—and some computers—can do computations can be very impressive, but fundamentally, the brain is fairly simple. Do the following problem by speaking aloud as you work through the process of getting the answer.

Multiply: $2\times3\times4$

Did you say, "2×3 is 6 and 6×4 is 24"? Or maybe you said, "4×3 is 12 and 12×2 is 24." Then there is the rare person who will say, "2×4 is 8 and 8×3 is 24." But in each case, do you see how the brain handles just two numbers at a time? Did you also notice that different approaches can lead to the same answer?

Addition and multiplication are much more flexible than subtraction and division in allowing for attaining answers using different approaches. For example, $2+3$ gives the same answer as $3+2$. In the same way, $2\times3=3\times2$. The ***commutative property***, which means that the operation can be performed in either order, exists for addition and for multiplication, but not for subtraction or for division. (Note that $2-3$ is not the same as $3-2$ and that $2\div3$ does not equal $3\div2$.) When operating with more than two numbers, we must decide which two to begin with. In mathematics, we often use grouping symbols to show the reader how a problem is done, and the most common grouping symbols are parentheses. Did you say 2×3 first or 3×4 first when you did the multiplication problem earlier? As we have seen, with multiplication we have some flexibility when determining the answer. Using grouping symbols, we see that $(2 \times 3) \times 4 = 2 \times (3 \times 4)$, with the problem in parentheses computed first. The **_associative property_** for addition and for multiplication states that the manner in which we group the initial numbers to be computed does not change the outcome. Verify for yourself that the associative property does _not_ work for subtraction or division by computing the values of $(12 - 5) - 6$ and $12 - (5 - 6)$ and then the values of $(40 \div 4) \div 5$ and $40 \div (4 \div 5)$.

Identities

The **_identity_** element for addition and subtraction is 0. Add 0 to any number or subtract 0 from any number, and the result is identical to the number with which you started. The identity element for multiplication and division is 1. The identity elements will prove crucial when you begin to solve equations. Getting back to the identities is also very important. Which number, when added to 6, gives the identity additive 0? Which number, when added to -7, gives the identity additive 0? You know that the answers are -6 and 7, respectively. Thus we say that the **_additive inverse_** of 6 is -6 and the additive inverse of 7 is -7.

ESSENTIAL

You may have called additive inverse numbers opposites in the past, but this may lead to confusion because the word _opposite_ can have a variety of interpretations. From now on, you should use the term _additive inverse_ so that you will become accustomed to it.

What number, when multiplied by 6, gives the multiplicative identity 1? What number, when multiplied by -7, gives the multiplicative identity 1? The answers are 1/6 and -1/7, respectively. We say that **_multiplicative inverse_** of 6 is 1/6 and the multiplicative inverse of -7 is -1/7. In the past you may have called these numbers reciprocals. Because the word _reciprocal_ has only this one meaning in mathematics, you may continue to use it, but be sure you understand that it has the same meaning as _multiplicative inverse_.

Distributive Operation of Multiplication over Division

The last of the properties to be examined in this section is the **distributive operation of multiplication over division**. How is the distributive property used? Compute 6 (7+3). According to the order of operations, do the work in the parentheses first: 7+3=10. You next do the problem $6 \times 10 = 60$. Note, however, that $(6 \times 7) + (6 \times 3) = 42 + 18 = 60$ is equivalent to the original problem.

RULE

Although this is frequently called the distributive property, be sure to remember that it is multiplication that is being distributed over addition.

An important example of the distributive property is the problem -6(-8+8). When we add first, the problem becomes -6(0)=0. When we apply the distributive property, the problem becomes (-6)(-8)+(-6)(8). (-6)(-8)+(-48) must equal 0. Thus (-6)(-8) must equal 48, which reinforces the rule that the product of two negative numbers is a positive number.

Evaluating Algebraic Expressions

A **variable** is a symbol used to represent a number. In your earlier education, you may have seen expressions that included a square or a rectangle in which you were to enter a number. Because this notation tends to be cumbersome, it is traditional to use letters to represent unknown quantities. Sentence fragments using variables are called algebraic expressions. When a relational verb such as $=$, $<$, or $>$ is added to the statement, the expression becomes a sentence. $2(l+w)$ and bh are algebraic expressions. Expressions are evaluated using the order of operations. For example, evaluating $2(l+w)$ with $l=10$ and $w=4$ yields $2(10+4)=2(14)=28$.

Evaluate each of the following with $a=4$, $b=-9$, $c=12$, and $k=-2$.

1. $5a - b(c+k^2) \div (b+1)$

Substitute the values. $5(4) - (-9)(12+(-2)^2) \div (-9+1)$

Parentheses first.　　$5(4) - (-9)(12 + 4) \div (-8)$

　　　　　　　　　　$5(4) - (-9)(16) \div (-8)$

Multiply/divide.　　$20 + 144 \div 8$

　　　　　　　　　　$20 + 18$

　　　　　　　　　　38

2.　　　b^2

We know that -9 squared is 81, because $(-9) * (-9)$ is 81. But if you are using a calculator and are not careful, you will get the incorrect answer -81. The entry in the calculator -9^2 involves the order of operations. First the number 9 is squared (exponents), and then this result is multiplied by -1.

ESSENTIAL

When using your calculator, include parentheses around negative numbers to avoid getting an incorrect answer by violating the order of operations.

3.　　　$\dfrac{b^2 + c^2}{7a + k - 1} + 5c - k(a - b)$

Substitute the values.　$\dfrac{(-9)^2 + (12)^2}{7(4) + (-2) - 1} + 5(12) - (-2)(4 - (-9))$

Parentheses first, remembering that the numerator and denominator are essentially grouped together.

$$\frac{81 + 144}{25} + 5(12) - (-2)(13)$$

$$\frac{225}{25} + 5(12) - (-2)(13)$$

Multiply/divide.	$9 + 60 - (\text{-}26)$
Add/subtract.	$69 + 26$
	95

Combining Algebraic Expressions

What do you get when you add 5 apples and 4 apples? 9 apples. What do you get when you add 5 apples and 4 oranges? 9 pieces of fruit.

Objects that have the same characteristic (apples, students, inches) can be added to and subtracted from each other, and their common characteristic can be maintained: 5 apples $+ 4$ apples $= 9$ apples; 5 inches $+ 4$ inches $= 9$ inches; 5 students $+ 4$ students $= 9$ students. When the characteristics are mixed, however, you *may* be able to create a common characteristic—such as that shared by those 9 pieces of fruit—but many times you cannot. For example, adding 5 apples and 4 students does not yield any kind of meaningful answer.

The same principle applies when adding and subtracting algebraic expressions. $5x + 4x = 9x$ makes sense if you think of x as the common characteristic, whereas $5x + 4w$ consists of different items. (If you knew the values of x and w, you could certainly make the substitutions and evaluate the expression, but this would not be combining algebraic terms.)

ESSENTIAL

A minus sign in front of a grouped set of numbers is tricky. Remember that you are really multiplying the entire group by -1.

Expressions may contain more than one variable and can be simplified as much as is possible. For example, $5x + 4w + 9x - 8w - 3x - 2w$ can be simplified by combining the terms in x and the terms in w separately and writing an algebraic expression as an answer. $5x + 9x - 3x = 11x$ and $4w - 8w - 2w = \text{-}6w$. Therefore, $5x + 4w + 9x - 8w - 3x - 2w$ can be simplified to $11x - 6w$.

Care must be taken when exponents are involved. $5x + 4x^2$ both show the variable x, but you must remember that x^2 means that x is used as a factor twice. Consequently, x and x^2 are different characteristics and cannot be combined. You can simplify $4x + 7x^2 + 11x - 5x^2$ to $15x + 2x^2$, but no further.

Simplify: $5a + 12b + 9ab - 4a + 7b - 3ab$

There are three variable expressions in this problem: a, b, and ab. Combining the common terms $5a - 4a + 12b + 7b + 9ab - 3ab$ yields $a + 19b + 6ab$. (The answer could be written as $1a + 19b + 6ab$, but according to the multiplicative identity, $1a$ is the same as a, so the 1 is not written.)

Simplify: $6(7a - 2b) - 5(3a - 4b)$

Because subtracting 5 is the same as adding -5, let's rewrite the problem as

$6(7a - 2b) + (-5)(3a - 4b)$

When we use the distributive property, this becomes $42a - 12b + (-15)a + 20b$, which equals $27a + 8b$. This method of doing the example is a bit cumbersome. A different approach is to group the terms that follow the subtraction into a single term, $6(7a - 2b) - [5(3a - 4b)]$. Apply the distributive property to get $42a - 12b - [15a - 20b]$. **Distribute the negative sign through the grouping symbols** to get $42a - 12b - 15a + 20b = 27a + 8b$.

Rules of Exponents

Variable expressions can be added or subtracted if they contain the exact same variable (characteristic). However, no such restrictions exist for multiplication and division. If the numbers being multiplied or divided have the same base, the rules of exponents can be applied to simplify the result. For example, $(x^4)(x^5) = x^9$ because a term with four factors of x is being multiplied by another term with five factors of x. This means that the product will have nine factors of x—hence the answer x^9. What is the value of $(y^7)(y^3)$? A value with seven factors of y multiplied by another three factors of y means that the product must have ten factors of y, or $(y^7)(y^3)=y^{10}$. In general, if m and n represent positive integers, then $(x^m)(x^n)=x^{m+n}$. That is, when multiplying terms that have exponents with a common base, keep the base and add the exponents.

Any number divided by itself is 1. (Division by zero is not defined, so you must exclude 0 from consideration.) Consequently, when k^8 is divided by k^3, the quotient is k^5 because the three factors of k in the denominator divide into three of the eight factors of k in the numerator, leaving five factors of k in the numerator. That is, $\dfrac{k^8}{k^3}=k^5$, and it is usually said that the three factors of k in the denominator *cancel* three factors of k in the numerator. In general, if m and n represent integers, then $\dfrac{x^m}{x^n}=x^{m-n}$. Calculate $\dfrac{k^{12}}{k^9}$. The nine factors of k in the denominator cancel nine factors of k in the numerator, so the result is k^3.

There are two other scenarios that need to be considered. What happens when the exponent in the denominator is larger than the exponent in the numerator? And what happens when the exponents in both the numerator and the denominator are the same? If the exponent in the denominator is larger, as in the problem $\dfrac{k^9}{k^{12}}$, then the three factors of k that remain will be in the denominator. That is, $\dfrac{k^9}{k^{12}}=\dfrac{1}{k^3}$. If you use the subtraction rule from above, you get $\dfrac{k^9}{k^{12}}=k^{-3}$. Therefore, negative exponents are an indication that one must take a reciprocal. That is, $w^{-4}=\dfrac{1}{w^4}$ and $z^{-8}=\dfrac{1}{z^8}$.

If the exponents in the numerator and denominator are the same, as in $\dfrac{k^9}{k^9}$, then we apply our knowledge that any number (except zero) divided by itself is 1, so $\dfrac{k^9}{k^9}=1$. Using the subtraction rule, $\dfrac{k^9}{k^9}=k^0$. Therefore, by definition, any number (except zero) raised to the zero power is equal to 1.

Calculate $(4x^3)(5x^5)$. This is the multiplication of four numbers: 4, x^3, 5, and x^5. Because the only operation is multiplication, we can rearrange the order of the factors through the commutative property to get $4*5*x^3*x^5$.

Since $4*5=20$ and $x^3*x^5=x^8$, the answer to the problem $(4x^3)(5x^5)$ is $20x^8$.

Calculate $(x^3)^4$. The exponent 4 indicates that the base should be written four times. That is, $(x^3)^4=x^3*x^3*x^3*x^3$. Each of these terms consists of three factors of x, for a total of twelve values of x being multiplied together. That is, $(x^3)^4=x^3*x^3*x^3*x^3=x^{12}$. In general, $(x^n)^m=x^{nm}$. That is, when a term being raised to a power is also raised to a power, keep the base and multiply the exponents.

Calculate $(3x^2)^4$. This problem is a combination of the two problems just completed. $(3x^2)^4=(3x^2)*(3x^2)*(3x^2)*(3x^2)$. Rearranging these factors yields $3*3*3*3*x^2*x^2*x^2*x^2=81x^8$. In other words, $(3x^2)^4=3^4*(x^2)^4=81x^8$.

Calculate $\left(\dfrac{8x^5}{4x^2}\right)^3$. Simplify what is inside the parentheses first to get $(2x^3)^3=2^3(x^3)^3=8x^9$.

Calculate $(4x^3)(5y^2)$. Rearrange the factors to $4*5*x^3*y^2$ to get the answer $20x^3y^2$.

ALERT

Although unlike terms cannot be added to yield a single item, they can be multiplied and divided because in those operations, you are not counting common characteristics.

Calculate $\left(\dfrac{8x^6y^3}{6x^2y^4}\right)^2$, and write your answer with positive exponents.

Simplify what is inside the parentheses. $\left(\dfrac{4x^4}{3y}\right)^2=\dfrac{4^2\,(x^4)^2}{3^2\,y^2}=\dfrac{16x^8}{9y^2}$

Exercises for Chapter 1

Classify each of the numbers in Exercises 1–4 as a counting number (N), a whole number (W), an integer (Z), a rational number (Q), an irrational number (I), or a real number (R). Include all the sets to which each number belongs.

1. 12

2. -7

3. $\dfrac{-2}{3}$

4. $\sqrt{8}$

Perform each computation first with pencil and paper. Then check your answer with a calculator.

5. $-9 + (-10)$

6. $-9 - (-10)$

7. $-9 * (-10)$

8. $-8 - 7$

9. $12 - (-3)$

10. $5 - 9$

11. $\dfrac{-20}{5}$

12. $\dfrac{-12}{-3}$

Evaluate each of the following expressions, using $a=6$, $b=3$, $c=\text{-}5$, and $d=\text{-}2$. First do the problem with pencil and paper, and then check your answer with a calculator.

13. $b+c$

14. ab

15. cd

16. c^2

17. b^3

18. d^3

19. $5a-10d$

20. $4c+7b$

21. $a(b+c)$

22. $d(c-b)$

23. $c(d-b)^2$

24. $\dfrac{b-c}{d-a}$

25. $8a-\dfrac{b-2d}{a+c}+c^2$

26. $\dfrac{4a+8b}{cd+2}-\dfrac{d-b}{c}$

Simplify each of the following algebraic expressions. (Use positive exponents in your answers.)

27. $12x + 9x - 18x$

28. $11y + 8w - 17y + 5w$

29. $8x + 7x^2 - 3x - 9x^2$

30. $19y^2 + 8y^3 - 7y - 9y^2 - 10y^3$

31. $3(4f + 3g) + 2(3f - 5g)$

32. $5(4q + 3r) - 4(2q - 5r)$

33. $(7x)(9x)$

34. $(4y)(5y^2)$

35. $(6x^2)^3$

36. $(9p^2)(2p)^3$

37. $\dfrac{27m^7}{18m^5}$

38. $\dfrac{50h^4k^3}{75h^2k^7}$

39. $\dfrac{(8v^3)^2}{(4v^2)^3}$

40. $\left(\dfrac{8v^3}{6v^2}\right)^3$

CHAPTER 2

Working the Process Backward: Solving Linear Open Sentences

Mathematics is used in the fields of business and science to make decisions, discover relationships, and make sense of the world. In this chapter, you will work with the next building block: solving linear equations and inequalities as pure mathematical statements. Learning to solve linear problems is an important step in learning to solve equations that arise in more involved problems.

One-Step Linear Equations

What number, when added to 7, gives an answer of 12? Given a little thought, you realize that subtracting 7 from 12 gives you the answer 5, and this is the number in question. If asked to determine which number, when added to 147, gives an answer of 321, you may find that you cannot determine the number immediately, but you know that when you subtract 147 from 321, you will arrive at the correct answer. After doing a few of these problems, whenever you read, "What number, when added to . . . ," your thought process will go to "I need to subtract these numbers to get the answer." The only thing to deal with after that is how cumbersome the numbers are. Are they as basic as $12-7$? Or are they larger numbers that require a pencil and paper or a calculator? (After all, who wants to subtract 12,345,678 from 98,765,432 mentally?)

What number, when multiplied by 6, gives an answer of 42? This is very similar to the previous problems, except that rather than subtracting, you will divide 42 by 6 to get the answer 7. The operations addition and subtraction and the operations multiplication and division are called **opposite operations** because they reverse results. For example, add 7 to a number. Want to get back to the original number? Subtract 7. Divide a number by 12. Want to get back to the original number? Multiply by 12. People who solve mathematical problems apply this concept all the time.

For a mathematician, the phrase "What number" is way too much to write. Why use ten letters when one will do? Mathematicians use a symbol, or **variable**, to represent this unknown value. The variable they choose is usually a letter of the alphabet, and that letter is generally x (because, after all, x marks the spot). The question of what number, when added to 7, gives 12 as an answer would thus be written as $x+7=12$. Similarly, "What number, when added to 147, gives 321?" could be written as $n+147=321$. (Some people prefer to use n (number) for the number in question, but you can use x if you so choose.) Accordingly, "What number, when multiplied by 6, yields 42?" translates to $6n=42$.

How would you solve each of the following equations?

$$x+98=400 \qquad y-79=36 \qquad 5m=80 \qquad \frac{z}{8}=19$$

$$n+8\frac{5}{6}=15\frac{2}{5} \qquad k-4\frac{2}{3}=8\frac{1}{6} \qquad \frac{15}{4}q=\frac{75}{16} \qquad \frac{z}{5.6}=9.41$$

We hope you took a moment to look at the problems and realized that the equations in the first column are addition problems. Therefore, you will subtract to find the variable. The equations in the second column are both subtraction questions, so you will add to find the solution to the problem. The third column has multiplication questions. Therefore, you will divide to get the correct values of the variable. And the last column has division questions, so you will multiply by the divisors to solve the problem.

Solving the equations involving fractions with a calculator may be a bit tricky. A mixed number such as 8 and 5/6 really means $8 + 5/6$, so we need to be careful when subtracting. Place parentheses around fractions when using them in a calculator to ensure that the order of operations is being maintained. To solve the equation $n + 8\frac{5}{6} = 15\frac{2}{5}$, subtract $8\frac{5}{6}$ from both sides of the equation to get $n = 15\frac{2}{5} - 8\frac{5}{6}$. On your calculator, it will look like this:

CALCULATOR SCREENS

TI 83-84	TI Nspire	
$(15 + 2/5) - (8 + 5/6)$	$15 + \frac{2}{5} - (8 + \frac{5}{6})$	$\frac{197}{30}$
6.56666667		

With the TI 83-84 series, press the MATH key and choose the FRAC option to convert your answer to a fraction.

$\frac{197}{30}$ is bigger than 6. If you would like to rewrite your answer as a mixed number, subtract 6 from the result and write the difference as a fraction.

To solve the equation $\frac{15}{4}q = \frac{75}{16}$, divide both sides of the equation by $\frac{15}{4}$ to get $q = \frac{75}{16} \div \frac{15}{4}$. The following display shows the computations with and without the use of parentheses. Which is correct?

DIVISION OF FRACTIONS

TI 83-84	TI Nspire	
75/16/15/4	$\dfrac{\dfrac{75}{16}}{\dfrac{15}{4}}$	$\dfrac{5}{64}$
.078125		
(75/16)/(15/4)	$\dfrac{\dfrac{75}{16}}{\dfrac{15}{4}}$	$\dfrac{5}{4}$
1.25		

Solving Multistep Linear Equations

If 7 is added to 8 times some number, the answer is 79. What is the number?

You might think to yourself, "If 7 is added to get 79, then 8 times the number I am looking for must be 72. Since $8*9$ is 72, the number I am looking for is 9." And you would be completely correct.

When we translate this problem, the phrase "8 times some number" becomes $8n$, where n is the number in question, and the phrase "7 is added to" becomes $8n+7$. The phrase "the answer is 79" finishes the sentence, and the whole equation is $8n+7=79$. The order of operations requires that you take the unknown number, multiply by 8, and add 7. In solving the equation, as the logic of the paragraph above shows, you remove the last item first. That is, because adding 7 was the last operation performed on the left side of the equation, removing the 7 by subtraction is the first needed to solve the problem and then divide this result by 8.

SOLVE: $8n+7=79$

Description	Action
Subtract 7.	$8n=72$
Divide by 8.	$n=9$

That is, you follow the order of operations in reverse order.

SOLVE: $18p + 873 = 1359$

Description	Action
Subtract 873.	$18p = 486$
Divide by 18.	$p = 27$

Here is a different-looking problem:

Solve: $945 - 12k = 705$

You could add $12k$ to both sides to get an equation that is similar to what you have already seen. You could also treat the subtraction problem as an addition of a negative number, in which case the problem becomes

Description	Action
Rewrite the equation with the variable term listed first.	$-12k + 945 = 705$
Subtract 945.	$-12k = -240$
Divide by -12.	$k = 20$

SOLVE: $14(3w + 19) = 980$

Description	Action
Apply the distributive property.	$42w + 266 = 980$
Subtract 266.	$42w = 714$
Divide by 42.	$w = 17$

There are also problems in which the variable appears more than once. Sometimes the variables are on the same side of the equation, and other times they are on opposite sides.

SOLVE: $15m + 84 + 7m - 17 = 265$

Description	Action
Combine like terms on the left-hand side of the equation.	$22m + 67 = 265$
Subtract 67.	$22m = 198$
Divide by 22.	$m = 9$

RULE

When the variable appears more than once in the same problem, the first step is always to gather these terms together into a single term.

When the equation contains variables on each side of the equation, a reasonable first step is to gather like terms on the same side of the equation. Here is an example:

SOLVE: $12x + 87 = 9x + 135$

Description	Action
Subtract 9x from both sides of the equation.	$3x + 87 = 135$
Subtract 87.	$3x = 48$
Divide by 3.	$x = 16$

A slightly different problem is to solve $\frac{12}{v} = 15$. You can argue that this problem would look much nicer if there were not any fractions in it. Multiplying both sides of the equation by the denominator v will remove the fraction from the problem.

SOLVE: $\frac{12}{v} = 15$

Description	Action
Multiply by v.	$12 = 15v$
Divide by 15.	$4/5 = v$

Solving $\dfrac{2x}{3}+\dfrac{x}{6}=10$ is another slightly different problem, because the fractions on the left-hand side of the equation are being added. Multiplying both sides of the equation by the common denominator will enable you to remove the fractions.

SOLVE: $\dfrac{2x}{3}+\dfrac{x}{6}=10$

Description	Action
Multiply by 6.	$6\left(\dfrac{2x}{3}+\dfrac{x}{6}\right)=10*6$
Use the distributive property.	$6\left(\dfrac{2x}{3}\right)+6\left(\dfrac{x}{6}\right)=60$
Simplify.	$4x+x=60$
Gather like terms.	$5x=60$
Divide by 5.	$x=12$

Solving Linear Inequalities

If you were asked, "Which number is larger: 10 or 30?" you would probably ask yourself whether the questioner had lost his mind before you confidently gave the answer 30. But you might not be as quick to respond if you were asked, "Which number is larger: -10 or -30?" Actually, there are a number of different ways to explain the concept of larger (or smaller), but for the purposes of solving linear inequalities, it is best to think of the larger number as the number that is farther to the right on the number line. In this case, then, we would say that -10 is greater than -30. In mathematical notation, -10 > -30. (You might also think of temperatures. As cold as it is, a temperature of -10 is still warmer than a temperature of -30.)

Solving linear inequalities is essentially the same as solving linear equations, with one very important difference. Multiplying, or dividing, both sides of an inequality by a negative number forces the direction of the inequality to be reversed. Think about the inequality $2x > 8$. What values of x will make this true? Some of the numbers that will make it true are 4.2, 5, 8.5, 29, and 112. In fact, after dividing both sides by 2, you can see that

$x > 4$ (that is, any number larger than 4) will solve the inequality. In a similar manner, the solution to the inequality $2x > -8$ will include -2, -1, 0, 2.1, 9, and 112. Again, dividing both sides of the inequality by 2, you get $x > -4$, which means that any number greater than -4 will solve the problem.

The solution to $-2x > 8$ requires a little more thought. $(-2)(-5) = 10$, so -5 is a solution. But $(-2)(-3) = 6$, and $6 < 8$, so 6 is not a solution. $(-2)(-4.5) = 9$, so -4.5 is a solution. In fact, if you choose any number less than -4, the product of that number and -2 will be larger than 8. If you divide both sides of the inequality $-2x > 8$ by -2, the result is $x < -4$, which means that any number less than -4 will solve the problem.

Graphing Solutions of Linear Inequalities

There is a subtle difference between the solution of $2x > 8$ and the solution of $2x \geq 8$. (Remember that the notation $\geq$ means "is greater than or equal to.") In the second case, 4 is a solution of the inequality, but in the first case, 4 is not a solution. When one is graphing the solution on the number line, how does one distinguish between these cases? You cannot start the solution for the first case at 5, because numbers such as 4.2, 4.5, 4.8762398 are part of the solution. In fact, you can get as close to 4 as you like, just so long as the number is larger than 4. An open circle is used to represent that you can get as close to the endpoint as you like but you cannot include the endpoint as part of the solution. In the second case, a closed circle is used as an endpoint to indicate that the endpoint is part of the solution.

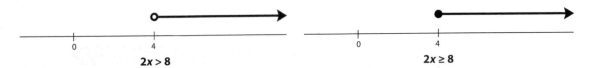

$2x > 8$ $2x \geq 8$

SOLVE $5x + 9 > 37$, AND GRAPH THE SOLUTION ON A NUMBER LINE.

Description	Action	Graph
Subtract 9 from both sides of the inequality.	$5x > 28$	
Divide by 5.	$x > 5.6$	(number line with open circle at 5.6, 0 marked) $x > 5.6$

SOLVE $9x - 23 \leq 3x - 12$, AND GRAPH THE SOLUTION ON A NUMBER LINE.

Description	Action	Graph
Subtract $3x$ from both sides, and add 23 to both sides.	$6x \leq 11$	
Divide by 6.	$x \leq 11/6$	$x \leq 11/6$

SOLVE $4n - 17 \geq 7n - 11$, AND GRAPH THE SOLUTION ON A NUMBER LINE.

Description	Action	Graph
Subtract $4n$ from both sides, and add 11 to both sides.	$-6 \geq 3n$	
Divide by 3.	$-2 \geq n$	$n \leq -2$

Did you notice that something was different about this problem? The variable is on the right-hand side of the inequality, not the left-hand side. Stop and think about what this means. The final inequality reads, "-2 is greater than or equal to all the solutions to this problem." Do you see that another way of writing this sentence is "All the solutions to this problem are less than or equal to -2"? That is, the inequalities $-2 \geq n$ and $n \leq -2$ represent the same set of numbers.

SOLVE $7m - 273 < 24m - 140$, AND GRAPH THE SOLUTION ON A NUMBER LINE.

Description	Action	Graph
Subtract $7m$ from both sides, and add 140 to both sides.	$-133 < 17m$	
Divide by 17.	$-19 < m$	$m > -19$

As you just saw, this is the same as saying that $m > -19$. When we are graphing the solution to this problem, it is not reasonable to sketch all the way to -19 on the number line counting by one. It is commonplace to show only two points on the number line, 0 and the endpoint.

Ratio and Proportion

Ratios have countless applications in our world. Maps and graphs used in newspapers and magazines must show a scale so that the reader can gauge distances and sizes accurately. Architects draw scale models of the buildings they plan to build and also use blueprints to show the construction workers how big various parts of the building should be. Clearly, the architect cannot build a model the size of the building itself to show to her customer!

From these few examples, you should see that ratios can be used to relate different measures. A map may use a scale of 1 inch : 100 miles. That is, 1 inch on the map corresponds to 100 miles in reality. An architect may use a scale of 1 cm : 3 m, which means that 1 centimeter on the blueprint corresponds to 3 meters in the building. Other ratios that you may have seen include 65 miles per hour, $10 per hour, $3 per gallon, and "Buy 2, get 1 free!"

A **proportion** results when two ratios are set equal to each other. 65 miles : 1 hour is the same as 130 miles: 2 hours or 650 miles : 10 hours. That is, $65{:}1 = 130{:}2 = 650{:}10$. Another way to write this is $\dfrac{65}{1} = \dfrac{130}{2} = \dfrac{650}{10}$. Proportions are very useful to determine an unknown quantity. For example, when reading a map, you may discover by measurement that two particular points on the map are 4.35 centimeters (cm) apart. The scale for the map reads 1 cm = 75 km (kilometers). What is the distance, d, between the points on the map in reality? Setting up a proportion, you get $\dfrac{1}{75} = \dfrac{4.35}{d}$. Cross-multiplying (multiplying both sides of the equation by the common denominator 75d) yields $d = (75)(4.35) = 326.25$ km. The two points are 326.35 km apart in reality.

SOLVE: $\dfrac{5x+9}{7}=\dfrac{3x-11}{4}$

Description	Action
Cross-multiply.	$4(5x+9)=7(3x-11)$
Distribute.	$20x+36=21x-77$
Combine like terms.	$113=x$

Applied Linear Equations

Mathematics is applied, problems are not given as equations but are stated in a paragraph. One of the most important steps in problem solving has to happen immediately — you need to determine what is being sought and to define that quantity with a variable. You then determine how the facts of the problem are related so that you can write a meaningful equation for the problem.

Example: Kate, a small business owner, knows that her cost, C, for doing business each month is $C=2.5n+12,000$, where n is the number of items she can produce each month.

Kate has \$15,000 available to spend for the month. How many items can she produce?

Kate can afford costs of 15,000 for next month, so $15,000=2.5n+12,000$. Solving this equation, she determines she can produce 1,200 items next month.

How much money will it cost her to produce 6,000 items in a month?

In this case, 6,000 is the number of items produced, so $C=2.5(6,000)+12,000=\$27,000$.

ESSENTIAL

When solving applications, it is very important to pay attention to the characteristic (or units) associated with each variable so that you will substitute values into the correct variables.

Example: The conversion between degrees Fahrenheit, F, and degrees Celsius, C, is $C=\dfrac{5}{9}(F-32)$.

FAHRENHEIT (F) AND CELSIUS (C) CONVERSION

Question/Description	Solution
What temperature in Celsius corresponds to 86°F?	$C = \dfrac{5}{9}(86 - 32) = \dfrac{5}{9}(54) = 30.$
What temperature in Fahrenheit corresponds to 86°C?	$86 = \dfrac{5}{9}(F - 32)$
Multiply by 9 to remove the fraction.	$774 = 5(F - 32)$
Apply the distributive property.	$774 = 5F - 160$
Add 160.	$934 = 5F$
Divide by 5.	$186.8 = F$

Motion is a very important topic in mathematics and in other areas. The basic formula regarding motion is that the rate of speed, r, multiplied by the time, t, gives the distance, d, traveled. That is, $d = rt$. You should remember that the units given in the rate should agree with the units of time and distance in the problem. Otherwise, you will have to convert one of these items to get a correct answer.

For example, a car traveling at 60 miles per hour for 4 hours travels a distance $d = (60)(4)$, for a total of 240 miles. A car traveling 60 miles per hour for 30 minutes travels a distance $d = (60)(0.5)$, for a total distance of 30 miles, because 30 minutes is the same as 0.5 hour.

Example: Suppose Drew and a friend leave his house at 10 o'clock in the morning on his way to a baseball game 200 miles away. Fifteen minutes after Drew leaves, his father finds the tickets for the game on the kitchen table. Drew's father knows the route Drew will take to the highway, and he knows that Drew will set his cruise control for 65 mph. Drew's father will need the same amount of time to get to the highway as Drew did, about 10 minutes. Once on the highway, the father sets his cruise control to 70 mph. Will the father reach Drew before he gets to the ball park?

It takes Drew and his father exactly the same amount of time to reach the highway. Therefore, you can think of the problem as beginning when Drew's father reaches the entrance to the highway. At this time, Drew has already traveled for 15 minutes at 65 mph, or 16.25 miles. Drew's distance from the entrance to the highway from that time forward is $d = 65t + 16.25$, while his father's position is explained by the equation $d = 70t$. When his father catches up to him, they will have traveled the same distance. This problem can be solved using the following relationship:

Drew's father's distance from the highway entrance $=$ Drew's distance from the highway entrance

$70t = 65t + 16.25$

$5t = 16.25$

$t = 3.25$

In 3.25 hours, Drew's father will have driven more than the 200 miles so he is already at the ball park. He might as well see if he can get another ticket to watch the game. (Or maybe Drew should buy him the ticket since he would not have gotten in without his father bringing the tickets he himself left on the table.)

Example: Kristen is visiting her grandparents. While there, she sees a jar full of coins on one of the bookshelves. Her grandfather tells her, "There is 30 dollars in that jar. There are some quarters, some dimes, and some nickels. There are 10 more nickels than quarters, and the number of dimes is 5 less than twice the number of quarters. If you can tell me how many of each kind of coin there are in the jar within the next 5 minutes, you can have the money."

Kristen asks her grandfather for a piece of paper and a pencil and begins the problem. The number of dimes and the number of nickels are related to the number of quarters, so she decides to let q represent the number of quarters in the jar. The number of nickels must be 10 more than the number of quarters, so there are $q + 10$ nickels in the jar. "Five less than twice the number of quarters," Kristen thinks to herself, "must mean that I need $2q$, twice the number of quarters, so the number of dimes must be 5 less than this, or $2q - 5$. Because I have statements for the number of each type of coin, now I can write an equation for the amount of money in the jar. I have to be consistent in how I record the amount of money. If I say a nickel is worth 5 cents, then I will need to write that the amount of money in the jar is 3,000 cents. If I want to say the amount of money in the jar is 30 dollars, then I will have to say that the nickel is worth 0.05 dollar. I can approach the problem either way, so I will choose to write the equations in cents. That way I will not have to use decimals."

Having made her decision, Kristen writes

$$25q + 10(2q - 5) + 5(q + 10) = 3,000$$

that is, the amount of money in quarters plus the amount in dimes plus the amount in nickels is 3,000 cents. Solving the equation, she writes

$$25q + 20q - 50 + 5q + 50 = 3,000$$

$$50q = 3,000$$

$$q = 60$$

Kristen then turns to her grandfather and announces, "There are 60 quarters, 115 dimes, and 70 nickels in the jar. Thank you, Grandpa!"

Applied Linear Inequalities

You want to have friends over on Saturday night to watch a movie. You look around the room where you will watch the movie and ask yourself, "How many people can comfortably fit into this room to watch the movie?" Suppose your answer is 10. Does this mean you have to invite 10 people over? No, you can invite no more than 10 people so that the room does not get over crowded. The number 10 in this case is a maximum bound on the number you can invite. (Because you cannot invite a negative number of people to watch the movie, your answer when written as an inequality would be $0 \leq$ number of people invited ≤ 10.)

Example: Brendon charges a $15 entrance fee to his art gallery. He knows from past experience that having n people visit on any given day will result in costs of $8n + 600$ for the day. How many people must come to the gallery each day for Brendon to make a profit? That is, how many people must come each day so that the money he takes in (his revenue) is greater than the money he is spending (his costs)?

Because Brendon charges $15 for each person who comes to the gallery and the number of people entering has been represented by n, his revenue for any day is $15n$. Brendon will make a profit when revenue is greater than costs.

Solve: $15n > 8n + 600$

Subtract $8n$ from both sides. $7n > 600$

Divide by 7. $n > 85.714$

Brendon needs at least 86 people to pay for entrance to his gallery each day so that he will make a profit.

Example: How many people need to pay for entrance to Brendon's gallery each day so that his daily profit is at least $300? Because profit is the money left over after paying the costs, profit = revenue − cost. The question asks for the profit to be *at least* $300, or profit ≥ 300, which is the same as

Revenue − cost ≥ 300

For Brendon's gallery, this problem becomes $15n - (8n + 600) \geq 300$.

Distribute the minus sign. $15n - 8n - 600 \geq 300$

Combine like terms by adding 600 to each side. $7n \geq 900$

Divide by 7. $n \geq 128.571$

Brendon needs at least 129 people to pay for entrance into his gallery so that his daily profit is at least $300.

Compound Inequalities

Who pays the full price when you go to a movie? Children under the age of 12 get a cheaper ticket as do senior citizens above the age of 62. If your age is between 12 and 62, that is, 12 < your age < 62, you pay full price. However, if your age is at most 12 or at least 62, your age ≤ 12 or your age ≥ 62, you can get a ticket to the movie at a lower price.

Compound inequalities can either include a continuous set of numbers, such as those who have to pay full price for a movie ticket, or unconnected sets, such as the ages of those patrons who get a discounted price for their movie ticket.

Type 1 diabetes is a serious health disorder in which the amount of glucose is outside a range of normal values. The American Diabetes Association states the normal pre-meal blood sugar level for people between the ages of thirteen and nineteen is 90–130 milligrams (mg) per deciliter (dL) of blood. That is, if the reading of a person's blood sugar level prior to a meal satisfies the inequality 90 ≤ blood sugar level ≤ 130, then that person is said to have a normal reading. If the blood sugar level reading is less than 90 (blood sugar level < 90), the person is hypoglycemic, and if the reading exceeds 130 (blood sugar level > 130), the person is hyperglycemic. Both of these conditions can lead to serious health issues. The following figure illustrates these numbers.

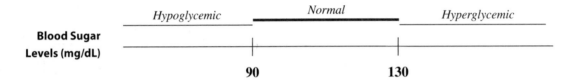

In one popular carnival game, a carnival worker tries to guess your age to within 3 years. If the person is incorrect, you win a prize. If Alysha, who is nineteen years old, pays to play this game, she will win if the carnival worker guesses her age to be less than sixteen (guess < 16) or greater than twenty-two (guess > 22). If the carnival worker selects a number from 16 to 22 (16 ≤ guess ≤ 22), then Alysha loses.

Take a moment to reread the inequalities in these two cases. $90 \leq$ blood sugar level ≤ 130 means that the blood sugar level is at least 90 *and* at most 130. Both of these conditions must be met. Alysha will win a prize at the carnival if the age guessed is less than 16 *or* greater than 22. The guessed age cannot meet both of these conditions at the same time, so they cannot be written in a single inequality. That is, the inequality $90 \leq$ blood sugar level *and* the inequality blood sugar level ≤ 130 can be condensed to the single statement $90 \leq$ blood sugar level ≤ 130, because that statement represents a continuous set of values. The conditions under which Alysha wins the carnival prize consist of two distinct sets of numbers.

Compound inequalities are solved in the same way as simple inequalities.

Solve, and graph the solution set: $5 \leq 3x - 7 \leq 17$

$5 \leq 3x - 7 \leq 17$ means that $5 \leq 3x - 7$ and $3x - 7 \leq 17$

$12 \leq 3x$ and $3x \leq 24$

$4 \leq x$ and $x \leq 8$, or $4 \leq x \leq 8$

Note that you had to perform the same steps (add 7; divide by 3) on both inequalities. Consequently, the problem could have been solved without separating the problem into two problems.

SOLVE AND GRAPH THE SOLUTION SET: $5 \leq 3x - 7 \leq 17$

Description	Action	Graph
Add 7.	$12 \leq 3x \leq 24$	
Divide by 3.	$4 \leq x \leq 8$	 4 8 **$4 \leq x \leq 8$**

SOLVE AND GRAPH THE SOLUTION SET: $-8 \leq 7 - 5x < 3$

Description	Action	Graph
Subtract 7.	$-15 \leq -5x < -4$	
Divide by -5 (remember to change the direction of the inequalities).	$3 \geq x > -4/5$	
Rewrite the inequality.	$-4/5 < x \leq 3$	 -4/5 3 **$-4/5 < x \leq 3$**

SOLVE: $-5 < 2x - 3$ and $3x - 4 \leq 5$

Description	Action
The binomials are different, so solve the inequalities separately.	
Add.	$-2 < 2x$ and $3x \leq 9$
Divide.	$-1 < x$ and $x < 3$
This the same as	$-1 < x < 3$.

SOLVE AND GRAPH THE SOLUTION SET: $2x + 11 < 9$ or $2x + 11 \geq 15$

Description	Action	Graph
Subtract 11.	$2x < -2$ or $2x \geq 4$	
Divide by 2.	$x < -1$ or $x \geq 2$	 $x \leq -1$ or $x \geq 2$

SOLVE AND GRAPH THE SOLUTION SET: $7 - 3x < 11$ or $7 - 3x > -5$

Description	Action	Graph
Subtract 7.	$-3x < 4$ or $-3x > -12$	
Divide by -3.	$x > -4/3$ or $x < 4$	 **Reals**

This is different from the other problems. Note that the solution covers the entire number line. For this problem, the solution is the set of all real numbers.

Exercises for Chapter 2

Solve each of the following equations. Fractions may be left as improper fractions, and decimals may be rounded to three decimal places.

1. $12n = 168$

2. $\dfrac{3}{7}x = 8\dfrac{3}{5}$

3. $4.9p = 11.466$

4. $y + 89 = 27$

5. $z - 3\dfrac{2}{9} = -2\dfrac{2}{3}$

6. $r - 13.4 = 21.7$

7. $3d + 17 = 44$

8. $23 - 11a = \text{-}43$

9. $4(2c - 13) = 56$

10. $2q - 3(8 - q) = 26$

11. $\dfrac{5}{8}n + 4\dfrac{3}{4} = 9\dfrac{5}{6}$

12. $18w + 29 = 7w - 92$

13. $3(4x + 9) - 2(7x - 12) = 81 - 3x$

14. $\dfrac{3}{4}(2x - 1) - \dfrac{2}{3}(3x + 4) = \dfrac{5}{6}(x + 7)$

15. $\dfrac{5x - 4}{7} = \dfrac{2x + 9}{3}$

16. $\dfrac{18}{3x - 9} = \dfrac{5}{x - 7}$

17. $\dfrac{2n + 9}{5n - 2} = \dfrac{1}{3}$

Solve each inequality, and graph the solution on a number line.

18. $4g - 13 < 12$

19. $7r + 19 \geq \text{-}16$

20. $17 - 7n < 52$

21. $\text{-}14 < 3x + 13 \leq 46$

22. $29 < 5 - 4x < 77$

23. $5x - 9 < 26$ or $3x + 10 > 76$

24. $14 - 11x < 80$ or $2x + 3 \leq 17$

25. Kristen's grandfather has a jar of coins that contains nickels, dimes, and quarters only. The total value of the coins is $48.50. The number of dimes is 20 more than twice the number of quarters, and there are 30 more nickels than dimes. How many of each type of coin is in the jar?

26. Sonya gets on the highway at 7 A.M. to make the 6-hour trip to see her client. She sets the cruise control at 65 mph. A little later, Sonya's colleague Alison realizes that she did not give Sonya all the paperwork Sonya will need for the meeting with the client. Alison puts the essential paperwork into her briefcase and gets on the highway, using the same exit, at 7:30 A.M. If Alison wants to catch up to Sonya in less than 2 hours, what is the slowest speed at which she can set her cruise control?

27. Xiao, Amy, and Jiang are 3 years apart in age, Xiao being the youngest and Jiang the oldest. If 30 more than the sum of Xiao's age and Amy's age is 3 times Jiang's age, is Amy old enough to vote in a general election in the United States? (*Hint:* Let a represent Amy's age.)

28. Juanita's expenses for her small pottery business total $15n + 320$ dollars per day when she has n pieces of pottery to sell. If Juanita charges $28 per piece, how many pieces must she sell each day to make a profit of at least $100 per day?

CHAPTER 3

Beyond the Line: Polynomial Expressions

Most problems in mathematics do not involve linear relationships but things that are curved. For example, the distance needed for a car to stop when the brakes are applied is related to the square of the speed of the car and the intensity of a light beam varies inversely with the square of the distance from the light source. In this chapter you will learn about the building blocks used to create these other kinds of problems, monomials and polynomials.

Adding and Subtracting Polynomials

To understand polynomials, you must first know about monomials. A *monomial* is a constant, a variable, or the product of constants and variables. Some examples are 7, x, $7x$, $8x^2$, $12x^2y$, and $3abc^3$. The *degree of a monomial* is the sum of the exponents in the problem for the variable terms. Thus x and $7x$ have degree 1 (they are therefore called linear or first-degree monomials); $8x^2$ has degree 2; $12x^2y$ has degree 3; and $3abc^3$ has degree 5. Terms that are added, subtracted, or divided are not monomials.

Polynomials consist of terms that are added to and subtracted from each other. The degree of a polynomial is the largest degree that appears in any of the monomials that make up the polynomial. $5x^2- 9x+8$ and $2x^3-4x^2-3x-7$ are polynomials. Each consists of terms that cannot be simplified into a single term (as one might be able to do with $4x+5x$) because they are not "like terms." Adding polynomials is a matter of adding like terms. That is, to add $5x^2-9x+8$ and $2x^3-4x^2-3x-7$, you can add $5x^2$ and $-4x^2$ to get x^2, $-9x$ and $-3x$ to get $-12x$, and 8 and -7 to get 1. Thus, $(5x^2-9x+8)+(2x^3-4x^2-3x-7)=2x^3+x^2-12x+1$.

Example: Add $8x^3+9x^2-11x-4$ and $3x^3-4x^2-5x-9$

$$(8x^3+9x^2-11x-4)+(3x^3-4x^2-5x-9)$$
$$=8x^3+3x^3+9x^2-4x^2-11x-5x-4-9$$
$$=11x^3+5x^2-16x-13$$

ALERT

Subtraction of polynomials often gives students trouble, because they forget to distribute the negative through the entire polynomial. Taking the time to write the parentheses and distributing the minus sign will help you avoid making such mistakes.

Subtraction of polynomials is just a bit trickier than subtraction of monomials, because you need to remember that the subtraction sign affects the entire polynomial that follows it. The procedure for subtracting polynomials is to consider the subtraction as multiplying the polynomial by -1, distribute the negative through the polynomial, and then add common terms. For example,

$$(8x^3 + 9x^2 - 11x - 4) - (3x^3 - 4x^2 - 5x - 9)$$
$$= 8x^3 + 9x^2 - 11x - 4 - 3x^3 + 4x^2 + 5x + 9$$

$$= 5x^3 + 13x^2 - 6x + 5$$

Polynomials need not consist of just one variable. $4x^3 + 5x^2y - 8xy^2 + 3y^3$ is a polynomial. In this case, each term of the polynomial has degree 3. The sum of this polynomial and $x^3 - 3x^2y - 7xy^2 + 5y^3$ can be simplified into a polynomial with four terms, because all of the terms that make up the polynomial are like terms:

$$(4x^3 + 5x^2y - 8xy^2 + 3y^3) + (x^3 - 3x^2y - 7xy^2 + 5y^3)$$
$$= 5x^3 + 2x^2y - 15xy^2 + 8y^3$$

However, the sum of $4x^3 + 5x^2y - 8xy^2 + 3y^3$ and $2x^3 + 6xy - xy^2 + 5y^4$ will contain more terms, because not all terms in the polynomial are like terms.

$$(4x^3 + 5x^2y - 8xy^2 + 3y^3) + (2x^3 + 6xy - xy^2 + 5y^4)$$
$$= 6x^3 + 4x^2y - 8xy^2 + 3y^3 + 6xy + 5y^4$$

Multiplying Polynomials by Monomials

Multiplying polynomials by monomials is an application of the distributive property. You know that $3(2x + 5) = 3(2x) + 3(5) = 6x + 15$. You also know that $(3x)(2x) = 6x^2$. Using these principles together reveals that $3x(2x + 5) = (3x)(2x) + (3x)(5) = 6x^2 + 15x$. Extending the distributive property allows you to expand $3x(7x^2 + 2x + 5) = (3x)(7x^2) + (3x)(2x) + (3x)(5) = 21x^3 + 6x^2 + 15x$.

The distributive property can also be used with multiple variables.

$$3xy^2(5x^2y - 4xy - 2xy^2) = (3xy^2)(5x^2y) - (3xy^2)(4xy) - (3xy^2)(2xy^2)$$
$$= 15x^3y^3 - 12x^2y^3 - 6x^2y^4$$

Multiplying Polynomials by Polynomials

Multiplying polynomials is an extension of the distributive property. Consider for the moment $A = 4x + 3$. Applying the distributive property to $A(2x - 5)$, you get $(A)(2x) - (A)(5) = 2xA - 5A$. Replacing the A with $4x + 3$, this result becomes $2x(4x + 3) - 5(4x + 3)$, which equals $(2x)(4x) + (2x)(3) + (-5)(4x) + (-5)(3) = 8x^2 + 6x - 20x - 15$. Combining like terms, you end with $8x^2 - 14x - 15$.

What does this look like without using A to represent the left factor? To multiply $(4x + 3)(2x - 5)$, you first distribute the binomial $(4x + 3)$ through $2x - 5$.

$$(4x + 3)\ (2x - 5) = (4x + 3)(2x) + (4x + 3)(\text{-}5)$$

Apply the distributive property to each of these results to get $8x^2 + 6x - 20x - 15$, and combine like terms to get the final answer.

Example: $(9n - 8)(7n - 3) = (9n - 8)(7n) + (9n - 8)(\text{-}3)$

$$= (9n)(7n) + (\text{-}8)(7n) + (9n)(\text{-}3) + (\text{-}8)(\text{-}3)$$

$$= 63n^2 - 56n - 27n + 24 = 63n^2 - 83n + 24$$

Example: $(3w - 4)(5w^2 + 7w - 8)$
$$= (3w - 4)(5w^2) + (3w - 4)(7w) + (3w - 4)(\text{-}8)$$

$$= (3w)(5w^2) + (\text{-}4)(5w^2) + (3w)(7w) + (\text{-}4)(7w) + (3w)(\text{-}8) + (\text{-}4)(\text{-}8)$$

$$= 15w^3 - 20w^2 + 21w^2 - 28w - 24w + 32$$

$$= 15w^3 + w^2 - 52w + 32$$

Example: $(5x^2 + 3)(6x^2 - 4x + 9)$
$$= (5x^2 + 3)(6x^2) + (5x^2 + 3)(\text{-}4x) + (5x^2 + 3)(9)$$

$$= (5x^2)(6x^2) + (3)(6x^2) + (5x^2)(\text{-}4x) + (3)(\text{-}4x) + (5x^2)(9) + (3)(9)$$

$$= 30x^4 + 18x^2 - 20x^3 - 12x + 45x^2 + 27$$

$$= 30x^4 - 20x^3 + 63x^2 - 12x + 27$$

Example: $(4x+7y)(9x^2-12xy-4y^2)$
$=(4x+7y)\ (9x^2)+(4x+7y)\ (\text{-}12xy)+(4x+7y)(\text{-}4y^2)$

$=(4x)(9x^2)+(7y)(9x^2)+(4x)(\text{-}12xy)+(7y)(\text{-}12xy)+(4x)(\text{-}4y^2)+(7y)(\text{-}4y^2)$

$=36x^3+63x^2y-48x^2y-84xy^2-16xy^2-28y^3$

$=36x^3+15x^2y-100xy^2-28y^3$

Example: $(3x-5)^2=(3x-5)(3x-5)$

$=(3x)(3x-5)+(\text{-}5)(3x-5)$

$=9x^2-15x-15x+25$

$=9x^2-30x+25$

Example: $(3x-5)^3=(3x-5)(3x-5)^2$

We make use of the answer from the previous problem:

$=(3x-5)(9x^2-30x+25)$

$=(3x)(9x^2-30x+25)+(\text{-}5)(9x^2-30x+25)$

$=(3x)(9x^2)+(3x)(\text{-}30x)+(3x)(25)+(\text{-}5)(9x^2)+(\text{-}5)(\text{-}30x)+(\text{-}5)(25)$

$=27x^3-90x^2+75x-45x^2+150x-125$

$=27x^3-135x^2+225x-125$

The previous examples illustrate that multiplying polynomials is an application of the distributive property. You may have encountered the mnemonic FOIL (First Outside Inside Last) for multiplying binomials. As you have seen, the distributive property applies to all cases, not just to the case of multiplying binomials.

Important Product Formulas

There are five product formulas with which you should be familiar because you will use them with some regularity during your study of mathematics. The first of these is the formula for the **difference of squares**, $(a-b)(a+b)=a^2-b^2$.

Example: $(5m-7)(5m+7)=(5m)^2-7^2=25m^2-49$

Example: $(6-11a)(6+11a)=6^2-(11a)^2=36-121a^2$

Example: $\left(\frac{2}{3}a+\frac{5}{8}\right)\left(\frac{2}{3}a-\frac{5}{8}\right)=\left(\frac{2}{3}a\right)^2-\left(\frac{5}{8}\right)^2=\frac{4}{9}a^2-\frac{25}{64}$

There are two variations on the formula for the **square of a trinomial** that you should know:

$$(a+b)^2=a^2+2ab+b^2 \quad \text{and} \quad (a-b)^2=a^2-2ab+b^2$$

Example: $(2n+7)^2=(2n)^2+2(2n)(7)+(7)^2=4n^2+28n+49$

Example: $(3m+4n)^2=(3m)^2+2(3m)(4n)+(4n)^2$
$$=9m^2+24mn+16n^2$$

Example: $(7c-6)^2=(7c)^2-2(7c)(6)+(6)^2=49c^2-84c+36$

Example: $(8v-3u)^2=(8v)^2-2(8v)(3u)+(3u)^2=64v^2-48uv+9u^2$

In each case, the first and third terms are the squares of the terms in the binomial, and the middle term is twice the product of the terms in the binomial. The sign in front of the middle term agrees with the sign between the terms in the binomial.

The fourth formula is the **difference of cubes**.
$$(a-b)(a^2+ab+b^2)=a^3-b^3$$
The fifth formula is the **sum of cubes**. $(a+b)(a^2-ab+b^2)=a^3+b^3$

Example: $(2x-5)(4x^2+10x+25)=(2x)^3-(5)^3=8x^3-125$

Example: $(3t-5s)(9t^2+15st+25s^2)=(3t)^3-(5s)^3=27t^3-125s^3$

Example: $(8q+7r)(64q^2-56qr+49r^2)=(8q)^3+(7t)^3=512q^3+343r^3$

Example: $\left(\dfrac{5}{3}w+\dfrac{2}{5}\right)\left(\dfrac{25}{9}w^2-\dfrac{4}{3}w+\dfrac{4}{25}\right)=\left(\dfrac{5}{3}w\right)^3+\left(\dfrac{2}{5}\right)^3=\dfrac{125}{27}w^3+\dfrac{8}{125}$

Dividing Polynomials by a Monomial

Is 5 a factor of 37? The answer is no because when 37 is divided by 5, the remainder is not 0. In this section you will learn to divide polynomials by monomials and polynomials. You will do this as a way for checking factors and also as a skill that needs to be developed for future work. (This is not a terrific reason but, as you know, there are places where groundwork needs to be laid so that you can see the applications in a future setting.)

$\dfrac{25x^5}{5x}=5x^4$, $\dfrac{35x^3}{5x}=7x^2$, and $\dfrac{40x}{5x}=8$. Using these results, it is reasonable to conclude that $\dfrac{25x^5-35x^3-40x}{5x}=5x^4-7x^2-8$. When dividing a polynomial by a monomial, divide each term of the polynomial by the monomial.

Simplify: $\dfrac{72y^6+45y^4-36y^3}{12y^2}=\dfrac{72y^6}{12y^2}+\dfrac{45y^4}{12y^2}-\dfrac{36y^3}{12y^2}=6y^4+\dfrac{15y^2}{4}-3y$

Simplify: $\dfrac{96z^3-60z^2+72z}{24z^5}=\dfrac{96z^3}{24z^5}-\dfrac{60z^2}{24z^5}+\dfrac{72z}{24z^5}=\dfrac{4}{z^2}-\dfrac{5}{2z^3}+\dfrac{3}{z^4}$

Division of Polynomials

Division of polynomials is very similar to the long division you did in middle school. Consider the process of dividing 5249 by 321:

Description	Action
Estimate how many times 5 can be divided by 3 (once).	
Multiply 321 by 1, write the result below 524, and subtract.	$\begin{array}{r} 16 \\ 321\overline{)5249} \\ 321 \\ \hline 2039 \\ 1926 \\ \hline 113 \end{array}$
Bring down the next term.	
3 goes into 20 about 6 times, so write the 6 above the 9.	
Multiply 321 by 9, and subtract.	

Because 103 is less than the divisor 321, we have that the quotient is 16 and the remainder is 113. That is, $5249 \div 321$ is $16\dfrac{113}{321}$.

Dividing polynomials is a similar process, but there are places where you will need to be careful. Read through each of the examples that follow. After you have read each problem, copy it onto a piece of paper and see if you can arrive at the correct answer. Be sure to be careful when doing the subtraction.

DIVIDE $6x^3 - 7x^2 - 47x - 36$ by $3x + 4$.

Description	Action
Divide the leading term of the dividend, $6x^3 - 7x^2 - 47x - 36$, by the first term of the divisor, $3x$. $6x^3 \div 3x = 2x^2$. When writing out the division problem, write this result over the x^2 term.	$$\begin{array}{r} 2x^2 \hspace{3.5cm} \\ 3x+4\overline{)6x^3-7x^2-47x-36} \end{array}$$
Multiply $3x + 4$ by $2x^2$, subtract, and bring down the remaining terms. Divide $-15x^2$ by $3x$ to get $-5x$. Repeat the multiply-and-subtract process.	$$\begin{array}{r} 6x^3+8x^2 \hspace{2cm} \\ \hline -15x^2-47x-36 \end{array}$$
Divide $-27x$ by $3x$ to get -9. Repeat the multiply-and-subtract process.	$$\begin{array}{r} 2x^2-5x-9 \hspace{1.5cm} \\ 3x+4\overline{)6x^3-7x^2-47x-36} \\ 6x^3+8x^2 \hspace{2cm} \\ \hline -15x^2-47x-36 \\ -15x^2-20x \hspace{0.8cm} \\ \hline -27x-36 \\ -27x-36 \end{array}$$

We find that $6x^3 - 7x^2 - 47x - 36$ by $3x + 4 = 2x^2 - 5x - 9$.

DIVIDE $10w^4 - 15w^3 - 2w^2 - 5w + 12$ by $2w - 3$.

Description	Action
$10w^4 \div 2w = 5w^3$. The first two terms subtract to 0. Bring down the next terms. Continue the process.	$$\begin{array}{r} 5w^3 \quad\quad -w-4 \\ 2w-3\overline{)10w^4-15w^3-2w^2-5w+12} \\ \underline{10w^4-15w^3}\quad\quad\quad\quad\quad \\ -2w^2-5w+12 \\ \underline{-2w^2+3w}\quad\quad \\ -8w+12 \\ -8w+12 \end{array}$$

Finishing the problem, you see that $10w^4 - 15w^3 - 2w^2 - 5w + 12$ divided by $2w - 3$ is $5w^2 - w - 4$.

Divide $18w^4 - 12w^3 - 3w^2 - 10w + 15$ by $3w - 2$.

$$\begin{array}{r} 6w^3 \quad\quad\quad -w-4 \\ 3w-2\overline{)18w^4-12w^3-3w^2-10w+15} \\ \underline{18w^4-12w^3}\quad\quad\quad\quad\quad\quad \\ -3w^2-10w+15 \\ \underline{-3w^2+\ 2w}\quad\quad \\ -12w+15 \\ \underline{-12w+\ 8} \\ 7 \end{array}$$

This problem has a remainder, so the solution is $6w^3 - w - 4 + \dfrac{7}{3w-2}$.

Example: Divide $8r^3 - 4r^2 + 19$ by $2r + 3$.

Note that there is no term between $-4r^2$ and 19 containing an r. We insert a placeholder, using a 0 for a coefficient. The steps for solving the problem remain the same.

$$\require{enclose}
\begin{array}{r}
4r^2-8r+12 \\
2r+3 \enclose{longdiv}{8r^3-\;4r^2+\;0r+19} \\
\end{array}$$

$$
\begin{array}{r}
\underline{8r^3+12r^2} \\
-16r^2+\;0r+19 \\
\underline{-16r^2-24r} \\
24r+19 \\
\underline{24r+36} \\
-17 \\
\end{array}$$

$8r^3-4r^2+19$ divided by $2r+3$ is $4r^2-8r+12-\dfrac{17}{2r+3}$.

Divisors do not always have to be binomials.

Example: Divide $2w^5+3w^4-4w^3+2w^2+10w-17$ by w^2+w-2.

$$
\begin{array}{r}
2w^3+\;w^2-w+5 \\
w^2+w-2 \enclose{longdiv}{2w^5+3w^4-4w^3+2w^2+10w-17} \\
\end{array}$$

$$
\begin{array}{r}
\underline{2w^5+2w^4-4w^3} \\
w^4+0w^3+2w^2+10w-17 \\
\underline{w^4+\;w^3-2w^2} \\
-w^3+4w^2+10w-17 \\
\underline{-w^3-\;w^2+2w} \\
5w^2+8w-17 \\
\underline{5w^2+5w-10} \\
3w-7 \\
\end{array}$$

Therefore, $2w^5+3w^4-4w^3+2w^2+10w-17$ divided by w^2+w-2
is $2w^3+w^2-w+5+\dfrac{3w-7}{w^2+w-2}$.

Exercises for Chapter 3

1. Find the sum of $7x^2 + 9x - 4$ and $15x^2 - 12x + 3$.

2. Find the difference when $7x^2 + 9x - 4$ is subtracted from $15x^2 - 12x + 3$.

3. Subtract $9x^3 + 4x - 17$ from $11x^3 - 3x^2 + 4$.

Find the indicated products.

4. $(8x + 7)(3x - 2)$

5. $(2w - 9p)(3w + 2p)$

6. $(6x - 5)(3x - 4)$

7. $(5w + 7v)(3w + 2v)$

8. $(7x + 2)(7x - 2)$

9. $\left(\dfrac{2}{3}r + \dfrac{5}{7}\right)\left(\dfrac{2}{3}r - \dfrac{5}{7}\right)$

10. $(9x - 8)(12x^2 - 7x - 5)$

11. $(8c - 5d)(2c^2 - cd - 3d^2)$

12. $(4m + 7)^2$

13. $(3k - 4z)^2$

14. $(4m + 7)^3$

15. $(3k - 4z)^3$

16. $(8g - 3)(64g^2 + 24g + 9)$

17. $(10r+9q)(100r^2-90rq+81q^2)$

18. Divide $12x^2-7x-10$ by $3x+2$.

19. Divide $32z^2-76z+35$ by $4z-7$.

20. Divide $15q^2+19q+7$ by $5q+3$.

21. Divide $28w^3+32w^2+7w+3$ by $7w+8$.

22. Divide $12w^3+32w^2-27w-72$ by $3w+8$.

23. Divide $15d^4+16d^3-15d^2+35d-21$ by $5d-3$.

24. The length and width of a rectangle are represented by $9x+7$ and $3x+5$, respectively. Write expressions to represent the perimeter and the area of the rectangle in terms of x.

CHAPTER 4

Graphing Linear Relationships

In the early seventeenth century, René Descartes changed the course of mathematics by creating the coordinate system. Every high school student in the modern world is completely familiar with the *x*- and *y*-axes, but the Cartesian coordinate plane was the first development in mathematics that allowed for geometers to use numbers and algebraists to use pictures. Descartes was one of the giants that Isaac Newton referred to in his famous quotation: "If I have seen further than others, it is because I have stood on the shoulders of giants."

The Cartesian Coordinate Plane

The Cartesian coordinate plane consists of two perpendicular number lines and a scale on each of these lines so that it is possible to identify any point on the plane with an ordered pair of numbers, or coordinates. The intersection of the two perpendicular lines is called the **origin** and has the ordered pair $(0, 0)$. The position of a point in the plane is found by first identifying the horizontal position relative to the origin; motion to the right is positive, and motion to the left is negative. This number is called the **abscissa**. The vertical distance from the origin is then determined; distance above the origin is positive, and distance below is negative. This number is called the **ordinate**. The horizontal axis is traditionally referred to as the x-axis, and the vertical axis is called the y-axis. As you continue your study of mathematics, you will be wise to remember that the horizontal axis is used for the input values and the vertical axis for the output values.

The following figure shows a number of points on the plane. Point A has coordinates $(3, 2)$, indicating that to get to A from the origin one must move right 3 and up 2. B has coordinates $(2, 3)$; note the "order" in which the ordered pairs are given. The coordinates for the other points are C(-2, 4), D(-5, 2), E(-4, -3), F(-2, -5), G(2, -4), H(4, -2), J(5, 0), K(0, 6), L(-3, 0), and M(0, -6). (The label I was intentionally omitted.)

Coordinates in the Cartesian Plane

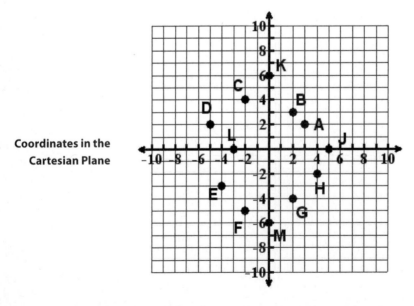

The two axes divide the plane into four regions called **quadrants**. Points A and B are in Quadrant I, points C and D are in Quadrant II, points E and F are in Quadrant III, and points G and H are in Quadrant IV. Note that in Quadrants I and IV the abscissas are positive, whereas in Quadrants II and III they are negative. The ordinates are positive in Quadrants I and II; they are negative in Quadrants III and IV. Points J and L lie on the x-axis and have a y-coordinate, or ordinate, of 0. Points K and M lie on the y-axis and have an x-coordinate, or abscissa, of 0. Points J, K, L, and M lie on an axis and (hence) are not in any of the quadrants.

Slope

You will sometimes hear people speak of a "straight line" when explaining a problem or giving directions to someone else. Do lines bend? In the classical case of line, the answer is no, so saying "straight line" is redundant. An interesting question then arises: "What makes a line straight?" One argument is that the angle through which the line moves is constant. Imagine Descartes explaining this application of numbers in a plane to his contemporaries. He might have said that the motion of the line is constant—that is, for every change in the horizontal motion, the corresponding change in the vertical motion must be proportional. He might also have taken advantage of everyone's picture of a mountain to explain this principle. There is some debate among the historians of mathematics about whether such conversations really happened, but the story does help to explain that the symbol chosen for slope is the letter m. The **slope** of a line is the ratio of the vertical change in the line (the rise) over a corresponding horizontal change (the run). Using the Greek uppercase letter Δ (delta) to stand for "change in," we can represent these two geometric interpretations by the formula $m = \dfrac{\text{rise}}{\text{run}} = \dfrac{\text{change in } y}{\text{change in } x} = \dfrac{\Delta y}{\Delta x}$. This discussion evolved into a formula for the slope of a line passing through two points with coordinates (x_1, y_1) and (x_2, y_2), where the subscripts 1 and 2 indicate the first ordered pair and the second ordered pair, respectively: $m = \dfrac{y_2 - y_1}{x_2 - x_1}$.

Example: Find the slope of the line joining: (a) point E(-4, -3) to point A(3, 2); (b) F(-2, -5) to B(2, 3); (c) E(-4, -3) to C(-2, 4); (d) M(0, -6) to H(4, -2)

a) Slope of $\overleftrightarrow{AE}$

$$m = \frac{2-(-3)}{3-(-4)} = \frac{5}{7}$$

(b) Slope of $\overleftrightarrow{FB}$

$$m = \frac{3-(-5)}{2-(-2)} = \frac{8}{4} = 2$$

(c) Slope of $\overleftrightarrow{EC}$

$$m = \frac{4-(-3)}{-2-(-4)} = \frac{7}{2}$$

(d) Slope of $\overleftrightarrow{MH}$

$$m = \frac{-2-(-6)}{4-0} = \frac{4}{4} = 1$$

In all four cases, the slopes of the lines are positive numbers. Use a ruler and a pencil to graph each of these lines. As you look at the graphs from left to right (which is the way all graphs in mathematics are analyzed), you should notice that all of the graphs are rising, or increasing.

Example: Find the slope of the line joining: (e) K(0, 6) to B(2, 3); (f) D(-5, 2) to L(-3, 0); (g) C(-2, 4) to G(2, -4); (h) A(3, 2) to J(5, 0)

(e) Slope of $\overleftrightarrow{KB}$

$$m = \frac{6-3}{0-2} = \frac{3}{-2} = \frac{-3}{2}$$

(f) Slope of $\overleftrightarrow{DL}$

$$m = \frac{0-2}{-3-(-5)} = \frac{-2}{2} = -1$$

(g) Slope of $\overleftrightarrow{CG}$

$$m = \frac{-4-4}{2-(-2)} = \frac{-8}{4} = -2$$

(h) Slope of $\overrightarrow{AJ}$

$$m = \frac{0-2}{5-3} = \frac{-2}{2} = -1$$

The slopes of the lines are negative numbers in these four problems. Draw the lines and observe that they all fall, or decrease, as they are examined from left to right. Also note that two of the lines, $\overleftrightarrow{DL}$ and $\overrightarrow{AJ}$, have the same slope and that, when drawn, these lines are parallel.

ESSENTIAL

Two lines will be parallel whenever their slopes are the same. Equal rates of change guarantee that the lines will not intersect.

Example: Find the slope of the line joining: (i) D(-5, 2) to A(3, 2); (j) L(-3, 0) to J(5, 0); (k) F(-2, -5) to C(-2, 4); (l) G(2, -4) to B(2, 3)

(i) Slope of $\overleftrightarrow{DA}$

$$m = \frac{2-2}{3-(\text{-}5)} = \frac{0}{8} = 0$$

(j) Slope of $\overleftrightarrow{JL}$

$$m = \frac{0-0}{\text{-}3-5} = \frac{0}{8} = 0$$

$\overleftrightarrow{DA}$ and $\overleftrightarrow{JL}$ are both horizontal lines; that is, they neither rise nor fall as they move from left to right.

(k) Slope of $\overleftrightarrow{CF}$

$$m = \frac{\text{-}5-4}{\text{-}2-(\text{-}2)} = \frac{\text{-}9}{0} = \text{undefined}$$

(l) Slope of $\overleftrightarrow{GB}$

$$m = \frac{3-(\text{-}4)}{2-2} = \frac{7}{0} = \text{undefined}$$

$\overleftrightarrow{CF}$ and $\overleftrightarrow{GB}$ are vertical lines; that is, they have no horizontal motion, and consequently their slopes must be undefined.

Graphing Lines

The Cartesian coordinate plane enables us to graph numerical relationships. For example, the equation $y = 3x + 2$ describes a relationship between the input values, x, and the output values, y. From a paper-and-pencil perspective, you could make a table of values for input-output pairs until you see a pattern.

x	$y = 3x + 2$
0	2
1	5
2	8
3	11

As you can see, these points lie on a line. What is the slope of this line? We can find out by arbitrarily choosing any two points. Say we choose the two points (0, 2) and (3, 11). We then calculate the slope to be $\frac{11-2}{3-0} = 3$, the same number that is the coefficient of x in the equation.

This equation can be graphed using graphing technology.

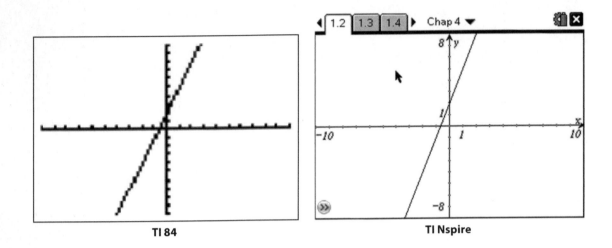

TI 84   TI Nspire

Example: Make a table of values for the points on the graph of $2x + 3y = 12$, and then sketch a graph for this equation.

x	$2x + 3y = 12$	y
0	$2(0) + 3y = 12$	4
1	$2(1) + 3y = 12$	10/3
2	$2(2) + 3y = 12$	8/3
3	$2(3) + 3y = 12$	2

Using the points (0, 4) and (3, 2) to compute the slope of this line, you get $m = \dfrac{2-4}{3-0} = \dfrac{-2}{3}$, which is the negative of the ratio of the coefficient of x to the coefficient of y.

You need to solve for y before entering the equation into your graphing technology. Subtracting $2x$ from both sides of the equation yields $3y = 12 - 2x$. Divide by 3, and you have $y = (12 - 2x)/3$.

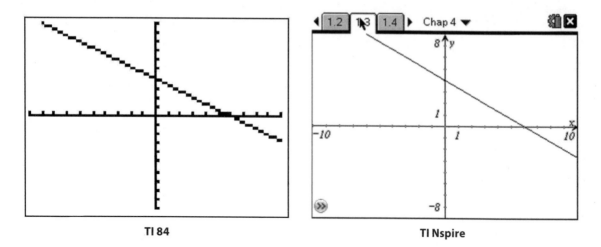

TI 84   **TI Nspire**

Example: Sketch the graph of a line that passes through the point (2, 3) and has slope$=3$.

Because the slope is an integer in this case, the denominator is a 1, and the slope can be interpreted as moving to the right 1 (change in x) and up 3 (change in y). Applying this motion from the point (2, 3) yields the point (3, 6). Plot both these points and draw the line that passes through them.

Example: Sketch the graph of the line that passes through the point (2, 3) and has slope$=$-2/5.

The negative sign in the slope is always associated with the change in y. From the point (2, 3), apply the change in x, 5, and the change in y, -2, to reach the next ordered pair (7, 1). Plot these points and draw the line between them.

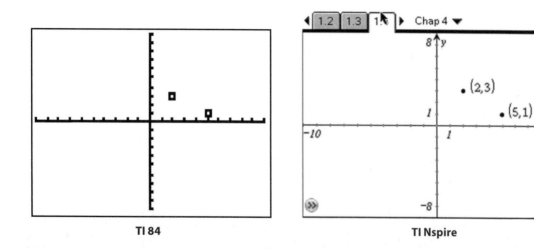

TI 84   **TI Nspire**

Intercepts

The points at which a graph crosses the coordinate axes are called the intercepts of the graph. The graph of the equation $y=3x+2$ crosses the y-axis at $y=2$—that is, when $x=0$, $y=2$. All points on the y-axis have an x-coordinate of 0, so the equation for the y-axis is $x=0$. In the same manner, all points on the x-axis have a y-coordinate of 0, so the equation for the x-axis is $y=0$. The x-intercept for the line $y=3x+2$ can be found by setting $y=0$ and solving for x to find that $x=$ -2/3.

The x-intercept for the graph with equation $2x+3y=12$ is found by setting $y=0$ and solving for x. The x-intercept is 6. The y-intercept for $2x+3y=0$ is found by setting $x=0$ and solving to find that $y=4$. You now know that the points (6, 0) and (0, 4) are two points on the graph of this equation, and you can use these two points to sketch a graph of the equation.

ESSENTIAL

The graph of an equation that has degree 1 will always be a line. For that reason, equations of degree 1 are called *linear equations*.

Example: Find the intercepts for the equation $5x-6y=24$, and use these results to find the slope of the line. Sketch a graph of the line.

x-intercept: $5x-6(0)=24$ yields $x=4.8$ $(4.8, 0)$

y-intercept: $5(0)-6y=24$ yields $y=$ -4 $(0, $ -4$)$

Slope: $m=\dfrac{0-(-4)}{4.8-0}=\dfrac{4}{4.8}=\dfrac{5}{6}$

Writing the Equation of a Line: Slope-Intercept Form

There are many different ways to write an equation for a line. You will learn three of these in this chapter: the slope-intercept form, the point-slope form, and the standard form. Given the equation $y=3x+2$, you know that the y-intercept for the graph is 2, and you can determine that the slope of the line is 3 by using the coordinates of two points that lie on the line. In fact, the two pieces of information immediately available to you as you read the equation are the slope and the y-intercept. Equations expressed in the form $y=mx+b$ are said to be in the ***slope-intercept form*** of the equation for a line for exactly that reason. You will not always be given such convenient information, but you can determine the equation for the line from the set of information you do have available.

Example: Write the slope-intercept equation for the line that passes through the point (-2, 5) and has a slope of 4.

Because the slope is 4, the equation of the line must be of the form $y=4x+b$. Knowing that $y=5$ when $x=-2$, you can substitute for x and y to determine the value of b.

$$y=4x+b$$

$$5=4(-2)+b$$

$$13=b$$

Equation: $y=4x+13$

Example: Write the equation of a line that passes through the points (3, -1) and (5, -7).

Although it is not given to you, you can determine the slope of the line: $m=\dfrac{-7-(-1)}{5-3}=\dfrac{-6}{2}=-3$. Given that the equation of the line must be of the form $y=-3x+b$, you can choose either of the given ordered pairs to substitute for x and y in this equation to find that $b=8$. The equation of the line is $y=-3x+8$.

Example: Write an equation for the line that passes through the point (3, -5) and is parallel to the line $y = 2x + 1$.

You know that the slope of the line you seek is 2, because parallel lines have the same slope. Substituting $x = 3$ and $y = -5$ into $y = 2x + b$ gives $b = -11$, and the equation is $y = 2x - 11$.

Example: Write an equation for the line that passes through the points (3, 5) and (7, 8).

As before, you can determine the slope from the formula; you find that $m = \dfrac{8-5}{7-3} = \dfrac{3}{4}$. Substituting the ordered pair (3, 5) for x and y into the equation $y = 3/4\ x + b$ gives $5 = 3/4\ (3) + b$, and $b = 11/4$. Therefore, the equation is $y = 3/4\ x + 11/4$.

Writing the Equation of a Line: Point-Slope Form

One of the advantages of using the point-slope form for a line is that two key pieces of information are given in the equation: the slope and a point (the point being the y-intercept). A disadvantage of using this formula is that when the slope or the coordinates of the points are especially easy to work with, the arithmetic for determining the slope and intercept gets sloppy. When we write an equation, the purpose of the variables x and y in the equation is to enable us to substitute values to determine whether a particular point is on the graph or to find a missing element of the ordered pair if one part is known. This can be used to write an equation for a line in what is called the ***point-slope form***.

RULE

A line that passes through the known point (x_1, y_1) with slope m has the equation $y - y_1 = m(x - x_1)$.

Example: Write an equation for the line that passes through the point (3, 5) and has slope 3/4.

Because the slope is 3/4, any point with coordinates (x, y) must satisfy the equation $\dfrac{y-5}{x-3} = \dfrac{3}{4}$. Do you realize this is all you need to do? You can check any point to determine whether it is on the line by substituting for x and y and seeing if the ratio on the left side of the equation is 3/4. This equation is usually modified by multiplying both sides of the equation by the denominator of the left-hand side so that the equation becomes $y-5=3/4 \ (x-3)$.

Example: Write an equation for the line that passes through the point (-4, 9) with slope -5/7.

$x_1=$-4 and $y_1=$9, so the equation is $y-9=$-5/7 $(x-(-4))$, which is simplified to $y-9=$-5/7 $(x+4)$.

Example: Determine the coordinates of the y-intercept and the x-intercept of the line with equation $y-9=$-5/7 $(x+4)$.

The y-intercept has an x-coordinate of 0. Substituting, $y-9=$-5/7 $(0+4)$ becomes $y-9=$-20/7, so $y=$43/7.

The x-intercept has a y-coordinate of 0. Substituting, $0-9=$-5/7 $(x+4)$ becomes -9=-5/7$(x+4)$. Multiply by -7/5 to get 63/5$=x+4$, or $x=$43/5.

Writing the Equation of a Line: Standard Form

The standard form of the equation for a line is not often written as an answer, but it is often given as a question. The ***standard form*** of the equation for a line is $Ax+By=C$, where A, B, and C are all integers. It is important for you to know how to graph a line given in standard form. It is less important to know how to write the equation in standard form. In fact, the standard form can usually be determined from the point-slope form without too much work.

Example: Sketch a graph of the line with equation $5x + 6y = 24$.

All you need to do to graph a line is plot two points. Finding the intercepts makes this easy to do. With $x = 0$, $y = 4$, and with $y = 0$, $x = 24/5$. Plot these points and sketch.

Example: Write an equation, in standard form, for the line that passes through the point (-4, 9) with slope -5/7.

Use the point-slope equation for the line in proportion form: $\dfrac{y-9}{x-(-4)} = \dfrac{-5}{7}$. Cross-multiply to get $7(y-9) = -5(x+4)$. Distribute to get $7y - 63 = -5x - 20$. Add $5x$ and 63 to each side of the equation to get the standard form $5x + 7y = 43$.

Graphing Linear Inequalities in Two Variables

If you were to roll one green die and one red die and to look at the numbers that showed on top, you would find that one of the following statements must be true: green > red, green = red, or green < red. Analogously, when you graph a line on a coordinate system, you create three regions on the plane: the line itself, the region above the line, and the region below the line.

All the points on the line satisfy the equation of the line. For example, the graph of the line $y = 2x + 1$ consists of all the points whose coordinates make the equation $y = 2x + 1$ a true statement. The point (-1, 4) is not on this line. Why? Substituting -1 for x and 4 for y yields the statement $4 = 2(-1) + 1$, or $4 = -1$. This is not a true statement. However, $4 > -1$ is a true statement, so the point (-1, 4) must lie on the graph of $y > 2x + 1$. What other points must be on this graph? (-1, 5), (-1, 6), (1, 7), . . . must be on this graph. Do you see that all these points are on the same side of the line $y = 2x + 1$?

RULE

When graphing an inequality, test one point that is not on the line itself. If the resulting inequality is true, shade everything on that side of the line. If the inequality is false, shade everything on the other side of the line.

When you sketch the graph of $y > 2x+1$, none of the points on the line $y = 2x+1$ can be included. Recall that when you are sketching on a number line, you distinguish between including and excluding the endpoint by using a filled circle (for inclusion of the endpoint) or an open circle (for exclusion of the endpoint). Similarly, when sketching an inequality with two variables, you will use a solid line for inclusion of the boundary and a dotted line for exclusion of the boundary.

Example: Sketch the graph of the solution to $y > 2x+1$.

The inequality "greater than" does not include an equals sign, so the boundary—those points for which y equals $2x+1$—will not be included in the graph. Plot some points for the line $y = 2x+1$, but then draw a dotted line through these points. The point (-1, 4) is not on the graph, so the correct region to shade will be the side which contains this point.

$y > 2x+1$

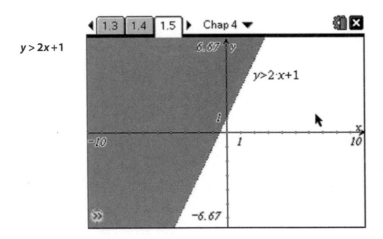

Example: Sketch the graph of the solution to $y \leq -3x+2$.

The inequality "less than or equal to" does include an equals sign, so the boundary line is included. Draw the line $y = -3x+2$. The point (0, 4) is not on this line. Putting the coordinates into the problem gives $4 \leq 2$, which is a false statement. The point (0, 4) is not part of the solution. Shade the part of the coordinate plane that lies on the opposite side of the boundary.

y ≤ -3x +2

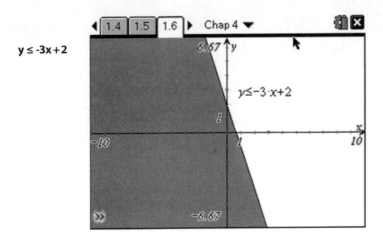

Exercises for Chapter 4

Plot each of the points on a piece of graph paper.

1. A(-3, 6)

2. B(1, -4)

3. C(5, 0)

4. D(5, 4)

5. E(-1, -7)

6. F(0, -3)

7. Identify the quadrants for the points A–F in Exercises 1–6.

Find the coordinates of the intercepts and the slope for each of the lines in Exercises 8–12.

8. $y = 6x - 5$

9. $y = 3/4\, x + 7$

10. $4x - 3y = 24$

11. $y - 2 = \dfrac{-2}{3}\,(x + 4)$

12. $y + 3 = \dfrac{8}{3}\,(x - 2)$

Write the slope-intercept equation for each line described in Exercises 13–16.

13. slope $=4$; point (-2, 5)

15. slope $=4/5$; point (-10, 3)

14. slope $=-7$; point (3, 0)

16. points (-2, -4) and (1, 5)

Write the point-slope equation for each line described in Exercises 17–20.

17. slope $=-2/3$; point (4, 7)

19. points (3, -5) and (-2, 7)

18. slope $=9/7$; point (-1, 4)

20. points (-5, 0) and (11, 7)

21. Write the standard form of the equation for the line in Exercise 19.

22. Write the standard form of the equation for the line in Exercise 20.

23. Find the intercepts of the line with equation $9x - 4y = 72$.

24. Find the slope of the line with equation $9x - 4y = 72$.

25. Sketch the solution to $y \geq 4x - 3$.

26. Sketch the solution to $2x - 3y > 6$.

CHAPTER 5

Functions

The relationship between two variables is an important topic in the study of mathematics. There is a difference between evaluating an expression such as $\sqrt{25}$ and solving the equation $x^2 = 25$. You will learn more about the difference in this chapter.

Relations and Functions

A ***relation*** is any set of ordered pairs. $A=\{(3, 4), (2, -1), (7, 2), (2, 0)\}$ is a relation (not a very interesting one, but a relation). The set of first elements (input values) in the relation is called the ***domain***, and the set of second elements (output values) is the ***range***. The domain of A is $\{2, 3, 7\}$ and the range is $\{-1, 0, 2, 4\}$. The number 2 is used twice as the input value for A, but it only has to be mentioned once in the domain statement.

Example: What are the domain and range of the relation $B=\{(0,-3), (5, 2), (4, -3), (2, -1), (9, 3)\}$?

The domain is the set of all first elements, so $D_B=\{0, 5, 4, 2\}$, and the range of B is $R_B=\{-3, 2, -1\}$. D and R represent the domain and range, and the subscript identifies the relation in question.

ALERT

A graphical test for whether a graph represents a function is the vertical-line test. If a vertical line can cross a graph more than once, the graph does not represent a function.

A special type of relation is a function. A ***function*** is a relation in which each element in the domain is paired with a unique element in the range. Of the two previous examples, A is not a function because the input value 2 is associated with the output values of -1 and 0. B is a function because each input value has a unique output. You may have noticed that -3 is used twice as an output in B, and that is acceptable. The requirement for a function is that the output for each input must be unique, not the other way around.

Function Notation

The instructor said, "Given $y=5x+7$, find the value of y when $x=8$. Given $y=-3x+5$, find the value of y when $x=2$. Now, find the value of y when $x=6$." Are you confused about which equation you should use? The first two directions from the instructor are very clear. The third is not, because both equations begin with $y =$. A notation has been created to help clarify a problem such as this and to also serve as a shorthand notation for problems such as the instructor gave. The function notation $f(x)$, which is read as "f of x," allows the user to label the equation and to announce the name of the independent (input) variable. $f(x)=5x+7$ and $y=5x+7$ say the same thing. "Given $y=5x+7$, find the value of y when $x=8$" is written "Find $f(8)$ if $f(x)=5x+7$." "Given $y=-3x+5$, find the value of y when $x=2$" becomes "Find $g(2)$ if $g(x)=-3x+5$." "Find the value of y when $x=6$" becomes "Find $f(6)$" or "Find $g(6)$", whichever equation the instructor had in mind, and the third command is now unambiguous. Thus $f(8)=47$ means that the point (8, 47) is a point on the graph of $y=f(x)$. And $g(2)=-1$ means that the point (2, -1) is a point on the graph of $y=g(x)$.

Example: Given $f(x)=3x-2$ and $g(x)=x^2+1$, and $p(x)=\dfrac{5x+3}{x-1}$.

Find $f(4), g(4), p(4), f(-2), g(-2), p(-2), f(z), g(z)$, and $p(z)$.

$$f(4)=3(4)-2=10 \qquad g(4)=(4)^2+1=17 \qquad p(4)=\frac{5(4)+3}{4-1}=\frac{23}{3}$$

$$f(-2)=3(-2)-2=-8 \qquad g(-2)=(-2)^2+1=5 \qquad p(-2)=\frac{5(-2)+3}{(-2)-1}=\frac{-7}{-3}=\frac{7}{3}$$

$$f(z)=3(z)-2=3z-2 \qquad g(z)=(z)^2+1=z^2+1 \qquad p(z)=\frac{5z+3}{z-1}$$

It is important to realize that the function notation is one of substitution. Whatever value is substituted for the x in $f(x)$ is also substituted for x in the rule that defines the function. Using $f(x)=3x-2$, as in the previous example, $f(4t-1)=3(4t-1)-2=12t-3-2=12t-5$.

What are the domains of $f(x)$, $g(x)$, and $p(x)$? As we have seen, the domain is the set of possible input values in a relation. Are there any values that must be avoided? At this point in your studies, you know that division by zero is not defined, so any number that causes the denominator of a fraction to equal zero must not be used. You also know that the square root of a negative number is not a real number, so such square roots must also be avoided. Thus the domain for $p(x)$ is $x \neq 1$. This implies that every other value of x is acceptable. $f(x)$ and $g(x)$ have no such restrictions, so the domain for each of these functions is the set of all real numbers.

Arithmetic with Functions

Businesses keep track of their costs and their revenues. The profit is computed when the costs are subtracted from the revenues. Most businesses project their costs, $C(x)$, and their revenues, $R(x)$, so that they can better prepare to compete with other businesses, attract investors, and, in general, be more successful. The business's profit, $P(x)$, is the difference between $R(x)$ and $C(x)$, $P(x) = R(x) - C(x)$.

Let $f(x) = 3x - 2$, $g(x) = x^2 + 1$, and $p(x) = \dfrac{5x+3}{x-1}$. These functions can be added, subtracted, multiplied, and divided together. $f(2) + g(4)$ can be computed by first finding $f(2) = 4$ and $g(4) = 17$. Therefore, $f(2) + g(4) = 4 + 17 = 21$.

$g(-2) * p(2)$ can be computed by finding $g(-2) = 5$ and $p(2) = 13$. Thus $g(-2) * p(2) = 5 * 13 = 65$.

$\dfrac{f(3) + g(1)}{p(3)}$ is computed by determining $f(3) = 7, g(1) = 2$, and $p(3) = 9$.

Therefore, $\dfrac{f(3) + g(1)}{p(3)} = \dfrac{7+2}{9} = 1$.

Mathematicians choose notation that is compact and unambiguous. (This is not always obvious when you are first learning the material, but as you get more accustomed to how the mathematics is written, the notation does make sense.) The input values for the previous two examples were different for each function. $f(2), g(4), f(3), g(1)$, and $p(2)$ do not lend themselves to a shorter approach. Suppose you are asked to compute $g(3) * p(3)$. The input values are the same, so all you really need to know is which functions and what operation are to be used. Because gp implies multiplication of the values of g and p, it should not be difficult to see that $g(3) * p(3)$ is written as $gp(3)$. Other examples are $g - f(4) = g(4) - f(4)$ and $\dfrac{p}{g}(2) = \dfrac{p(2)}{g(2)}$.

Example: Given $f(x) = 3x - 2, g(x) = x^2 + 1$, and $p(x) = \dfrac{5x+3}{x-1}$, find (a) $f+g(0)$; (b) $fp(2)$; (c) $\dfrac{g}{p}(-4)$

(a) $f(0)=-2$ and $g(0)=1$, so $f+g(0)=-2+1=-1$

(b) $f(2)=4$ and $p(2)=13$, so $fp(2)=(4)(13)=52$

(c) $g(-4)=17$ and $p(-4)=17/5$, so $\dfrac{g}{p}(-4)=\dfrac{17}{\frac{17}{5}}=5$

Composition of Functions

If $f(x)=3x-2$, then $f(3)=3(3)-2=7$. If $g(x)=x^2+1$, then $g(7)=7^2+1=50$. If $g(x)$ is evaluated with the answer from $f(x)$, as was done in the previous example, the notation will read $g(f(3))=50$. The function f is evaluated at 3, and this answer is used as the input value for g. Evaluating a function with the output of another function is called the **composition of functions**. $f(g(3))=f(3^2+1)=f(10)=3(10)-2=28$. Composition of functions is not necessarily commutative, so you must be careful about the order in which you compute the functions.

There is a second notation for composition that can be a bit more confusing. This notation comes from transformational geometry and uses a small, open circle. $f(g(x))=f \circ g(x)$. You may be thinking that $f \circ g$ is a comment about the weather, but it is not. Keep in mind that $f \circ g(x)$ is evaluated from right to left.

Important: In both notations $f(g(x))$ and $f \circ g(x)$, you first calculate with the rule that is closest to the input value.

Example: Given $f(x)=3x-2$, $g(x)=x^2+1$, and $p(x)=\dfrac{5x+3}{x-1}$, find: (a) $g(f(2))$; (b) $g \circ p(3)$; (c) $g \circ g(4)$; (d) $p \circ p(3)$

(a) $g(f(2))=g(4)=17$

(b) $g \circ p(3)=g(p(3))=g(9)=82$

(c) $g \circ g(4)=g(g(4))=g(17)=290$

(d) $p \circ p(3)=p(p(3))=p(9)=48/8=6$

Inverse Functions

Addition and subtraction are considered inverse operations because they reverse each other's process. The same is true of multiplication and division. Add 5 to a number. Subtract 5 to get back to the original number. Some functions, but not all functions, also have inverses. If B is the function $B = \{(0, -3), (5, 2), (4, -3), (2, -1), (9, 3)\}$, then the inverse of B will be the set of ordered pairs with the members of each pair given in reverse order. That is, to find the inverse of a function, interchange the x- and y-coordinates. For reasons that will be explained shortly, the notation for the inverse of B is B^{-1}. $B^{-1} = \{(-3, 0), (2, 5), (-3, 4), (-1, 2), (3, 9)\}$. Is B^{-1} a function? The answer is no, because the input value -3 is associated with two different output values.

RULE

To find the inverse of a function, interchange the x- and y-coordinates and solve for y.

What if the function is defined by a rule? For example, find the inverse of $f(x) = 5x + 7$. Remember that writing $f(x)$ is the same as writing y. Write the problem f: $y = 5x + 7$. Following the definition of inverse, interchange the x- and y-coordinates to get f^{-1}: $x = 5y + 7$. To complete the problem, solve this equation for y. $x - 7 = 5y$, so $y = f^{-1}(x) = \dfrac{x-7}{5}$. Graph each of these functions on your graphing calculator, along with the line $y = x$. Note that a geometric interpretation of finding the inverse is reflecting the graph of $f(x)$ across the line $y = x$.

RULE

The composition of a function f(x) and its inverse f⁻¹(x) will always be x.

Example: Find the inverse of $g(x) = \frac{-2}{3}x + 5$.

g: $y = \frac{-2}{3}x + 5$ becomes g^{-1}: $x = \frac{-2}{3}y + 5$

Subtract 5 and multiply by -3/2 to get $y = g^{-1}(x) = \frac{-3}{2}(x - 5)$

Example: Compute $g(g^{-1}(x))$ and $g^{-1}(g(x))$.

$g(g^{-1}(x)) = g(\frac{-3}{2}(x - 5)) = \frac{-2}{3}\left[\frac{-3}{2}(x - 5)\right] + 5 = 1(x - 5) + 5$

$= x - 5 + 5 = x$.

$g^{-1}(g(x)) = g^{-1}(\frac{-2}{3}x + 5) = \frac{-3}{2}(\frac{-2}{3}x + 5 - 5) = \frac{-3}{2}(\frac{-2}{3}x) = x$

Finally, why was the -1 in the exponent position chosen for the notation? After all, b^{-1} means $1/b$, the reciprocal of b. Using $g(x) = \frac{-2}{3}x + 5$, we find that $g(0) = 5$ and $g(6) = 1$. The ordered pairs (0, 5) and (6, 1) lie on the graph of g, whereas the ordered pairs (5, 0) and (1, 6) lie on the graph of g^{-1}. Notice that the slopes of the lines joining these points are -2/3 and -3/2, respectively, and that these numbers are reciprocals. No matter what the function is, these reciprocal relationships will exist, and that is how the notation came to be.

Exercises for Chapter 5

For Exercises 1 and 2: A relation is defined. (a) Determine the domain and range for the relation. (b) Determine whether the relation represents a function. (c) Find the inverse for the relation. (d) Determine whether the inverse represents a function.

1. $A = \{(3, 4), (5, 2), (0, 1), (1, 4), (-2, 5)\}$

2. $B = \{(5, 2), (-1, 3), (3, 5), (5, 1)\}$

For Exercises 3–19, let $f(x) = 3x + 8$, $g(x) = 2x - 3$, $k(x) = \dfrac{1}{2}x + 3$, and
$r(x) = \dfrac{3x - 4}{2x - 1}$.

Find

3. $f(4)$

4. $g(2)$

5. $k(1)$

6. $r(4)$

7. $f + g(3)$

8. $gk(-2)$

9. $\dfrac{k}{r}(-2)$

10. $r - f(1)$

11. $f(g(3))$

12. $g(f(3))$

13. $g(k(4))$

14. $k(g(4))$

15. $r(f(x))$

16. $g(k(x))$

17. $g(r(x))$

18. $f^{-1}(x)$

19. $k^{-1}(x)$

20. Find the domain of $r(x)$.

CHAPTER 6

Systems of Linear Equations

There are an infinite number of points on a line. Put two lines in the same plane, and one of the following scenarios must be true: The lines intersect in one point, the lines intersect in an infinite number of points, or the lines intersect in zero points. You will study three ways (graphically, algebraically, and with matrices) to determine which of these scenarios is true.

Solving Graphically

From a paper-and-pencil perspective, solving systems of linear equations algebraically works out well when the y-intercept is a "nice" number because the plotting of points will work out well. For example, if one of the equations is $y = -2/3\, x + 5$, you know that $(0, 5)$ is one point and that every time the graph moves to the right by a value of 3, it drops by a value of 2. You can plot the points $(3, 3)$ and $(1, 1)$ to graph this line. If the equation is given in standard form and both the slope and the y-intercept are fractions, such as $3x - 11y = 14$, then finding points with integer coefficients becomes much more challenging. Can you see that it will be difficult to find points to plot when the y-intercept at $14/11$ and the slope is $3/11$?

Graphing with technology enables us to solve these problems more conveniently. Lines whose equations are given in point-slope form are easily entered into the graphing calculator. Lines whose equations are given in standard form need to be transformed so that they can be entered into the equation editor of the graphing calculator.

Example: Solve the system of linear equations:
$$y = -2/3\, x + 5$$
$$3x - 4y = 14$$

Rewrite the second equation in standard form so that it reads "$y =$" but do not simplify the expression. $3x - 4y = 14$ becomes $-4y = 14 - 3x$ by subtraction, and then $y = (14 - 3x)/-4$ by division. The parentheses are critical to maintaining the validity of the equation. Graph each of these equations on the same set of axes. Use the intersect feature of the calculator to find the coordinates of the points of intersection.

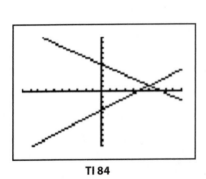

TI 84

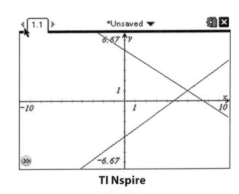

TI Nspire

FINDING POINTS OF INTERSECTION—TI 84

Description	Image
Calc: 2nd TRACE	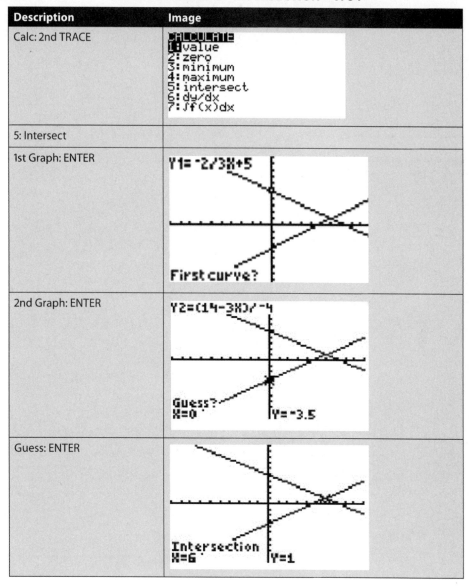 CALCULATE 1: value 2: zero 3: minimum 4: maximum 5: intersect 6: dy/dx 7: ∫f(x)dx
5: Intersect	
1st Graph: ENTER	Y1=-2/3X+5 First curve?
2nd Graph: ENTER	Y2=(14-3X)/-4 Guess? X=0 Y=-3.5
Guess: ENTER	Intersection X=6 Y=1

FINDING POINTS OF INTERSECTION—TI NSPIRE

Description	Image
Menu	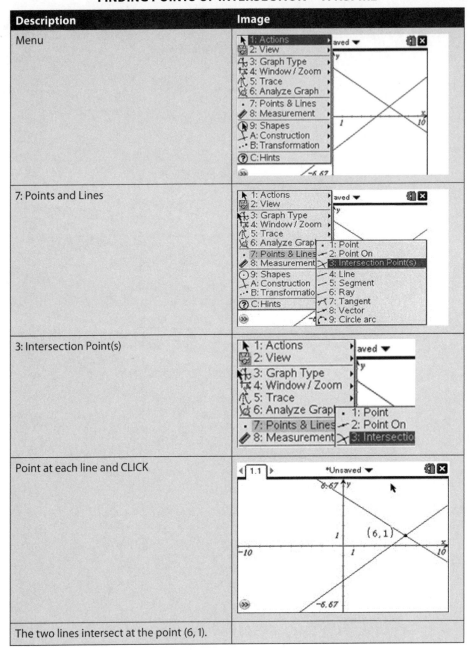
7: Points and Lines	
3: Intersection Point(s)	
Point at each line and CLICK	
The two lines intersect at the point (6, 1).	

Solve: $y = 5x - 7$

$y = 3x - 3$

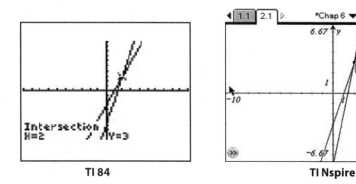

The two lines intersect at the point $(2, 3)$.

Solving Algebraically—Substitution

There are also a number of ways to solve systems of equations algebraically. In this section you will learn the process of substitution. This is most easily done when at least one of the equations is written in the slope-intercept form.

Solve: $y = 5x - 7$

$y = 3x - 3$

Substitute $3x - 3$ from the second equation (which is equal to y) for y in the first equation.

$3x - 3 = 5x - 7$

$4 = 2x$

$2 = x$

Substitute 2 for x in one of the original equations. For example, substitute into $y=5x-7$ to get $y=5(2)-7=3$.

The point of intersection for the two lines is $(2, 3)$.

Solve: $y=3x-4$

$$3x-5y=10$$

Substituting $3x-4$ from the first equation for y in the second equation, the problem becomes

$$3x-5(3x-4)=10$$

$$3x-15x+20=10$$

$$\text{-}12x=\text{-}10$$

$$x=5/6$$

Substituting 5/6 for x in the first equation, you get $y=3(5/6)-4=\text{-}3/2$. The two lines intersect at the point $(5/6, \text{-}3/2)$.

Solve: $2x+y=\text{-}5$

$$4x+5y=\text{-}7$$

Rewrite the first equation for y so that it reads $y=\text{-}2x-5$. Substitute this for y in the second equation to get

$$4x+5(\text{-}2x-5)=\text{-}7$$

$$4x-10x-25=\text{-}7$$

$$\text{-}6x=18$$

$$x=\text{-}3$$

Substitute $x=\text{-}3$ into the first equation to get $2(\text{-}3)+y=\text{-}5$. Solve for y, getting $y=1$. The two lines intersect at the point $(\text{-}3, 1)$.

Solving Algebraically—Elimination

The method of elimination (also known as the multiply-and-add process) is used when solving systems of linear equations wherein both equations are given in standard form. The advantage is that there is no need to work with a number of fractions, and the answer can always be found with a minimum number of steps.

RULE

When solving systems of equations by the elimination method, choose factors that will cause the coefficients attached to one of the variables to be negatives of one another.

Solve: $4x + 5y = -7$

$2x - y = 7$

Multiply both sides of the second equation by 5. The system now becomes

$4x + 5y = -7$ $\qquad\qquad$ $4x + 5y = -7$

$(2x - y = 7)\, 5$ $\qquad\qquad$ $\underline{10x - 5y = 35}$

Add the two equations. $14x \qquad = 28$

$$x = 2$$

Substitute $x = 2$ into the first equation. $4(2) + 5y = -7$. This becomes $5y = -15$, or $y = -3$. The two lines intersect at the point (2, -3).

Solve: $3x + 4y = 13$

$5x + 7y = 24$

Multiply the first equation by 5 and the second equation by -3 to eliminate the variable x. (Please note that multiplying the first equation by 7 and the second by -4 would enable you to eliminate the variable y.)

$$5(3x+4y=13)$$

$$15x+20y=65$$

$$-3(5x+7y=24)$$

$$\underline{-15x-21y=\text{-}72}$$

Add the two equations together.

$$-y=\text{-}7$$

$$y=7$$

Substitute $y=7$ into the first equation to get $3x+4(7)=13$. Solve to find $x=\text{-}5$. The two lines intersect at the point (-5, 7).

ALERT

If all variables drop out of the problem when you are adding, then the system represents the same line when the sum is true and represents parallel lines when the sum is false.

Solve: $4x+3y=5$

$$8x+6y=10$$

Multiply the first equation by -2 and add the equations.

$$-2(4x+3y=5) \qquad -8x-6y=\text{-}10$$

$$8x+6y=10 \qquad \underline{8x+6y=10}$$

$$0=0$$

Both variables dropped out of the problem. What does this mean about the system of equations? Look closely, and you will see that the second equation is twice the first equation. Use your graphing calculator to show that the two lines are the same line.

Applications of Linear Systems: A Building Block to Problem Solving

Application of linear equations had you set up one equation for the purpose of solving a problem about one quantity. In this section, you will solve problems about more than one entity. In order for you to do so, you will need to be able to write as many equations for the problem as there are quantities to find. You will need two equations to find two unknowns, three equations for three unknowns, and so on.

Example: Hillary visits her aunt. While there, she sees a jar with coins in it. Her aunt tells her that there are 200 coins in the jar, but only quarters and nickels. The total value of the money in the jar is $23.00. If Hillary can tell her aunt how many of each type of coin are in the jar, then she can have the money. How many quarters and how many nickels are in the jar?

There are two relationships in this problem to which Hillary needs to pay attention: There are 200 coins of only two types, and the value of these coins is $23.00.

Hillary reasons that an equation is needed for each of these relationships. She knows she can express the information she has about the number of coins by assigning the variable n to the number of nickels and the variable q to the number of quarters. She writes the following notes on her paper.

n: number of nickels

q: number of quarters

Number of coins	$n+q=200$
Value of coins	$5n+25q=2300$

Using this system of equations, Hillary multiplies both sides of the first equation by -5 to get

$-5n-5q=\text{-}1000$

$5n+25q=2300$

Adding the two equations together, Hillary gets $20q = 1300$, or $q = 65$. Therefore, $n = 135$. Checking her work, she computes $5(135) + 25(65) = 2300$, so she is confident that she can tell her aunt there are 135 nickels and 65 quarters in the jar. She gives her aunt a kiss on the cheek as her aunt hands her the money.

Example: Andrew's math class has 30 students in it. In a "Guys versus Girls" review for the unit test, he notices that that there are 4 more females in the class than males. How many students of each gender are in the class?

Letting f be the number of females in the class, and letting m be the number of males, Andrew sets up the following equations:

$$f + m = 30$$

$$f = m + 4$$

Do you see that because there are more females than males, Andrew added 4 to the number of males? Substituting for f in the first equation from the second equation, Andrew gets $m + 4 + m = 30$. When he solves the equation, Andrew can see that there are 13 males and 17 females in the class.

Example: Brenda is a freshman at college. Her Introduction to Psychology class has 295 students enrolled, and there are 33 more males than females. How many students of each gender are in the class?

Just as Andrew did in the previous question, Brenda can use the variables f and m to represent the numbers of students of each gender. The equations Brenda will write to solve her problem are

$$f + m = 295$$

$$m = f + 33$$

Substituting, and solving the equation $f + f + 33 = 295$, Brenda determines that there are 131 females and 164 males in her Introduction to Psychology class.

Example: Kate had $240 with which she can purchase blouses and shorts during the "No Sales Tax on Clothing Week" in her state. Blouses cost $15 each, and shorts cost $20. If Kate bought a total of 14 articles of clothing, how many blouses and how many shorts did she buy?

b: number of blouses bought

s: number of shorts bought

$$b+s=14$$

$$15b+20s=240$$

Multiplying both sides of the first equation by -15, you get

$$-15b-15s=-210$$

$$15b+20s=240$$

Adding these equations yields $5s=30$, so $s=6$. If Kate bought 6 shorts, then she must have purchased 8 blouses also.

Example: Have you ever noticed that it takes less time to fly from west to east than it does to fly from east to west? This is because the wind blows from west to east as a consequence of the spinning of the earth. The wind increases the speed of a plane that is flying east but slows the plane down as it flies west. When writing the equations in this section, keep in mind the basic formula for motion, Rate * Time = Distance.

Two planes flying parallel routes leave simultaneously. One plane flies west from London to New York City, while the other plane flies east from New York City to London. The air miles between London and New York City total 3500. The westbound plane makes the trip in 10 hours, and the eastbound plane makes the trip in 7 hours. Find the speed of the plane in still air and the speed of the wind.

Let p be the speed of the plane in still air.

Let w be the speed of the wind.

Westbound flight: $10(p-w)=3500$

Eastbound flight: $7(p+w)=3500$

Divide the first equation by 10 and the second equation by 7 to get the following equivalent system of equations:

$$p-w=350$$

$$p+w=500$$

Add these equations to get $2p=850$. Solving this equation yields $p=425$, so $w=75$. The speed of the plane in still air is 425 mph, and the speed of the wind is 75 mph.

Using Matrices to Solve Linear Systems

It is not uncommon for a car dealership to have multiple stores across a geographic region. Consider the case of the US Auto Import chain with stores in Yonkers, Croton-on-Harmon, Saratoga Springs, Syracuse, and Ithaca. The May inventory is taken of four of its best sellers: the Toyota Avalon, the Dodge Durango, the Jaguar S-type, and the BMW 530i sedan. Table 6.1 shows how many of each model are in stock at each location.

TABLE 6.1

Location	Avalon	Durango	S-Type	530i
Yonkers	25	5	15	25
Croton-on-Harmon	10	5	5	10
Saratoga Springs	12	10	3	12
Syracuse	8	15	6	15
Ithaca	20	10	10	15

The wholesale price of each model (that is, the price that US Auto Import paid for each model) is given in Table 6.2.

TABLE 6.2

Model	Wholesale Price
Avalon	$24,500
Durango	$26,800
S-Type	$42,000
530i	$43,500

Example 1: Compute the value of the inventory for the month of May at each of the five US Auto Import stores.

For each store, multiply the number of each model by the wholesale cost of the model.

Yonkers: $25(24,500) + 5(26,800) + 15(42,000) + 25(43,500)$
$= \$2,464,000$

Croton: $10(24,500) + 5(26,800) + 5(42,000) + 10(43,500)$
$= \$1,024,000$

Saratoga Springs: $12(24,500) + 10(26,800) + 3(42,000) + 12(43,500)$
$= \$1,210,000$

Syracuse: $8(24,500) + 15(26,800) + 6(42,000) + 15(43,500)$
$= \$1,502,500$

Ithaca: $20(24,500) + 10(26,800) + 10(42,000) + 15(43,500)$
$= \$1,830,500$

We will now use this example to illustrate the mathematical construct called a matrix. A ***matrix*** is a rectangular array of numbers. Ignoring the labels that are included to help read the tables, Table 6.1 has five rows and four columns. Rows are read horizontally and columns vertically. Table 6.2 has four rows and one column. The ***dimensions of a matrix*** are determined by the number of rows and the number of columns.

To enter a new matrix in TI 83+/84, the keystrokes are 2nd x^{-1}, left arrow, ENTER, and type the dimensions of the matrix. The dimensions of the matrix, 5 rows and 4 columns, are displayed on the top line of the screen as 5×4. The notation $A_{5,4}$ indicates that the matrix has dimensions 5×4.

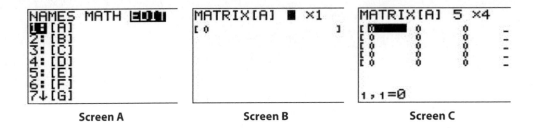

Screen A   Screen B   Screen C

Notice the number in the lower left-hand corner of the screen. The entry in column 1 row 1 is 0. For this problem, row 1 corresponds to the inventory in Yonkers. Enter 25 and press e.

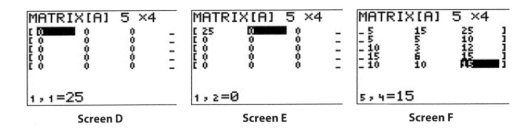

Screen D   Screen E   Screen F

As 25 is entered into the calculator, Screen D shows the display. After the ENTER key is struck, the calculator moves to the second column of row 1 and indicates that the current entry is 0. Finish entering the inventory values for matrix A. The limitations of the screen size and the size of the data prevent you from seeing all the data at once, but you can use the left and right arrows to scroll across the screen to be sure you have entered the data correctly (Screen F).

Edit matrix B to be 4×1 (4 rows, 1 column), and enter the values for the wholesale costs in a similar manner. Quit, and return to the home screen. Multiply matrices [A] and [B] by typing 2nd x⁻¹ ENTER x 2nd x⁻¹ 2 to get

$$[A] * [B] \begin{bmatrix} 2464000 \\ 1024000 \\ 1210000 \\ 1502500 \\ 1830500 \end{bmatrix}$$

These are the same values calculated in the solution to Example 1; they reflect the total value of the inventory for each store. This tells you the process in which multiplication of matrices occurs. Starting with row 1 of the left-hand factor, each number in row 1 is multiplied by a number from column 1 in the right-hand factor, and these products are added to give the result.

Matrix multiplication and the inverse of a matrix can be used to solve systems of linear equations. For instance, consider the following system:

$$3x+y=29$$

$$5x+3y=39$$

This is the equivalent of the matrix equation

$$\begin{bmatrix} 3 & 1 \\ 5 & 3 \end{bmatrix}\begin{bmatrix} x \\ y \end{bmatrix}=\begin{bmatrix} 29 \\ 39 \end{bmatrix}$$

Observe that the first matrix contains the same numbers as the coefficients from the system. The second matrix contains the variables of the system. The third system contains the constants of the linear equations. When you perform the matrix multiplication on the left, you see the same terms as appear on the left side of the system of equations.

$$\begin{bmatrix} 3x+y \\ 5x+3y \end{bmatrix}=\begin{bmatrix} 29 \\ 39 \end{bmatrix}$$

This matrix equation can be written as [A][X]=[B]. In Algebra 1, you would solve the equation $ax=b$ by dividing both sides of the equation by a to get $x=b/a$. However, you can't divide matrices. You could also solve the equation by multiplying both sides of the equation by the multiplicative inverse of a, $\frac{1}{a}ax=\frac{1}{a}b$, which gives the same result, $x=b/a$. Multiplication of matrices is *not* a commutative operation, so the order of operations must be adhered to carefully. For the purposes of solving the system of equations, you need to multiply with the inverse as the left factor. (If you try to multiply with the inverse as a right factor, the calculator will give you an error message.)

[A][X]=[B]

$[A]^{-1} [A][X]=[A]^{-1} [B]$

$[X]=[A]^{-1} [B]$

The solution to the system of equations is the point (12, -7).

$$\begin{bmatrix} x \\ y \end{bmatrix} = \begin{bmatrix} 3 & 1 \\ 5 & 3 \end{bmatrix}^{-1} \begin{bmatrix} 29 \\ 39 \end{bmatrix}$$

$$\begin{bmatrix} x \\ y \end{bmatrix} = \begin{bmatrix} 12 \\ -7 \end{bmatrix}$$

Solve: $5x+12y=-12$

$\qquad 3x+45y=18$

The matrix equation is

$$\begin{bmatrix} 5 & 12 \\ 3 & 45 \end{bmatrix} \begin{bmatrix} x \\ y \end{bmatrix} = \begin{bmatrix} -12 \\ 45 \end{bmatrix}$$

Multiply by the inverse of the coefficient matrix to get

$$\begin{bmatrix} x \\ y \end{bmatrix} = \begin{bmatrix} 5 & 12 \\ 3 & 45 \end{bmatrix}^{-1} \begin{bmatrix} -12 \\ 45 \end{bmatrix}$$

$$\begin{bmatrix} x \\ y \end{bmatrix} = \begin{bmatrix} -4 \\ 2/3 \end{bmatrix}$$

Linear Systems with More Than Two Variables

Systems of linear equations are not limited to two variables. In fact, you can have as many variables as you would like. In general, you should have as many equations as you have variables. Solving systems with more than three variables geometrically is very difficult, and when the number of variables is more than three, it is not physically possible. It is possible to use the elimination process to solve such systems. Using a system with three variables, the process is to pair one of the equations with each of the other two and then eliminate the same variable from each of these pairs.

Solve: $2x + 3y + 4z = 6$

$5x - 4y + 6z = 36$

$-3x - 8y + 12z = 7$

Match the first equation with the second and with the third, and arbitrarily choose to eliminate the variable x.

Solve: $2x + 3y + 4z = 6$ $2x + 3y + 4z = 6$

$5x - 4y + 6z = 36$ $-3x - 8y + 12z = 7$

Multiply each pair by appropriate factors.

$5(2x + 3y + 4z = 6)$ $3(2x + 3y + 4z = 6)$

$\underline{-2(5x - 4y + 6z = 36)}$ $\underline{2(-3x - 8y + 12z = 7)}$

$10x + 15y + 20z = 30$ $6x + 9y + 12z = 18$

$\underline{-10x + 8y - 12z = -72}$ $\underline{-6x - 16y + 24z = 14}$

$23y + 8z = -42$ $-7y + 36z = 32$

Match up these equations, multiplying by appropriate factors to eliminate the variable z.

$$9(23y+8z=\text{-}42) \qquad\qquad 207y+72z=\text{-}378$$

$$\underline{\text{-}2(\text{-}7y+36z=32)} \qquad\qquad \underline{14y-72z=\text{-}64}$$

$$221y=\text{-}442$$

$$y=\text{-}2$$

Substitute: $23(\text{-}2)+8z=\text{-}42$ becomes $8z=4$, so $z=1/2$.

Substitute into one of the original equations: $2x+3(\text{-}2)+4(1/2)=6$ becomes $2x=10$ and $x=5$.

Check these numbers in the other two original equations to ensure that the solution is correct. $5(5)-4(\text{-}2)+6(1/2)=25+8+3=36$ and $\text{-}3(5)-8(\text{-}2)+12(1/2)=\text{-}15+16+6=7$.

The solution to the system is the ordered triple (5, -2, 1/2).

The matrix solution to this problem requires much less writing. The matrix equation for the original problem is

$$\begin{bmatrix} 2 & 3 & 4 \\ 5 & \text{-}4 & 6 \\ \text{-}3 & \text{-}8 & 12 \end{bmatrix} \begin{bmatrix} x \\ y \\ z \end{bmatrix} = \begin{bmatrix} 6 \\ 36 \\ 7 \end{bmatrix}$$

The solution to this problem is

$$\begin{bmatrix} x \\ y \\ z \end{bmatrix} = \begin{bmatrix} 2 & 3 & 4 \\ 5 & \text{-}4 & 6 \\ \text{-}3 & \text{-}8 & 12 \end{bmatrix}^{-1} \begin{bmatrix} 6 \\ 36 \\ 7 \end{bmatrix} = \begin{bmatrix} 5 \\ \text{-}2 \\ 1/2 \end{bmatrix}$$

Solve:
$$9w+2x+3y-7z=\text{-}10$$
$$2w-3x+4y+3z=39$$
$$\text{-}5w+3x-y+6z=33$$
$$2w-10x+3y+z=0$$

Writing this as a matrix equation, you get

$$
\begin{bmatrix}
9 & 2 & 3 & -7 \\
2 & -3 & 4 & 3 \\
-5 & 3 & -1 & 6 \\
2 & -10 & 3 & 1
\end{bmatrix}
\begin{bmatrix}
w \\ x \\ y \\ z
\end{bmatrix}
=
\begin{bmatrix}
-10 \\ 39 \\ 33 \\ 0
\end{bmatrix}
$$

Multiply both sides of the equation by the inverse of the coefficient matrix to get

$$
\begin{bmatrix}
w \\ x \\ y \\ z
\end{bmatrix}
=
\begin{bmatrix}
9 & 2 & 3 & -7 \\
2 & -3 & 4 & 3 \\
-5 & 3 & -1 & 6 \\
2 & -10 & 3 & 1
\end{bmatrix}^{-1}
\begin{bmatrix}
-10 \\ 39 \\ 33 \\ 0
\end{bmatrix}
=
\begin{bmatrix}
-2 \\ 3 \\ 10 \\ 4
\end{bmatrix}
$$

The solution is the ordered 4-tuple (-2, 3, 10, 4). Funny word, isn't it?

Systems of Linear Inequalities

Solving systems of linear inequalities requires finding the set of points that meet the conditions of all the inequalities in the problem. Graphing more than one inequality on a coordinate system can get sloppy as you shade the multiple solutions. In the first two examples that follow you will see what it looks like to shade the solution for each of the inequalities. In the examples after that, the shading for each inequality is indicated with an arrow, and only the common solution is shown.

Determine the common solution for the linear inequalities, and state the coordinates of one point in the common solution.

Example: $y > 4x - 3$

$$3x + 5y \leq 12$$

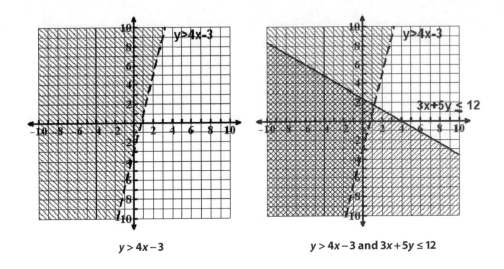

y > 4x - 3

y > 4x - 3 and 3x + 5y ≤ 12

One point in the common solution is (-2, 0).

Example: $2x - 3y > -6$

$$x \leq 5$$

$$y > 1$$

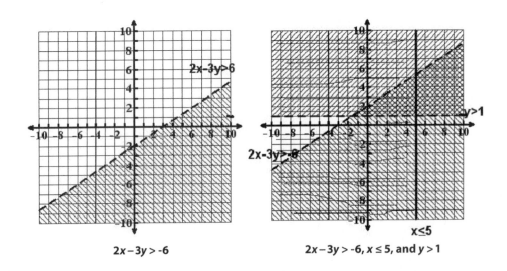

2x - 3y > -6

2x - 3y > -6, x ≤ 5, and y > 1

The common solution is the triangular region enclosed by the boundaries, and the point (4, 3) is in this common solution.

As you can see, the inequality with more than two boundaries is difficult to read because of all the regions that must be shown. Rather than shade the complete solution, it is customary to indicate the direction of the shaded region with an arrow for each of the boundaries and to shade only the common solution at the end of the problem. Using the same example, the solution would look like this:

2x − 3y > -6, x ≤ 5, and y > 1

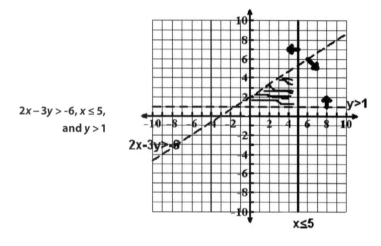

x≤5

Example: Sketch the common solution, and find the coordinates for a point in this solution.

$$2x + y < 9$$

$$y > 2x + 1$$

$$2x - 3y > 5$$

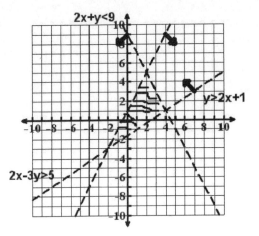

Common Solution

The point (2, 1) is in the common solution.

Exercises for Chapter 6

Solve each of the following systems of linear equations graphically.

1. $y = 3x + 17$

   $y = -2x - 3$

2. $y = 4x + 23$

   $5x + 6y = -7$

3. $y = -x$

   $6x + 3y = -15$

Solve each of the following systems of linear equations using the substitution method.

4. $y = 4x - 15$

   $6x - 5y = 33$

5. $y+2x=0$

   $16x-12y=15$

6. $y=7x-35.2$

   $5x+3y=16.6$

Solve each of the following systems of linear equations using the elimination method.

7. $2x+y=0$

   $3x+4y=8.2$

8. $14x+3y=13$

   $7x-9y=24$

9. $3x+4y=2$

   $5x+7y=39$

10. $3a+5b+4c=17$

    $2a-6b+7c=87$

    $8a+3b+2c=-19$

Solve using matrices.

11. $4x+5y=20$

    $7x+12y=61$

12. $3a+5b+4c=17$

    $2a-6b+7c=87$

    $8a+3b+2c=-19$

13. $7p + 9q + 6r = 1$

    $2p - 3q + 12r = 10$

    $10p + 7q - 4r = -2$

14. Sunny has been saving her change in a plastic jar that she had on her dresser. When the jar was filled, she found that she had 106 coins (dimes and quarters only) for a total of $19.30. How many of each type of coin did she have?

15. There are 424 students in the graduating class of South High School. One of the teachers noted that this is one of the most balanced classes she has ever seen, because there are only 6 more females than males in the class. How many females and how many males are in the class?

16. Professor Stockwell is looking at her book inventory. She determines that she will need to replace 23 books in all, for a total of $1325. If the textbooks cost $75 each and the answer books cost $25 each, how many of each type of book does she need?

17. Tim has been saving his change in a glass jar that he keeps in his room. When the jar was full, he found that he had 450 coins that totaled $62.15. He was amused to note that the number of dimes exceeded the number of nickels and quarters combined by 16, which happened to be his uniform number for the baseball team. How many coins of each type did Tim have in his jar?

18. A small jet made the 500-mile flight from Harrisburg, Pennsylvania, to South Bend, Indiana, and back on the same day under the same weather conditions. The westbound flight from Harrisburg took 2.5 hours, and the return trip took 2 hours. Determine the speed of the plane in still air and the speed of the wind.

Solve the following systems of linear inequalities graphically, and determine the coordinates of one point in the common solution.

19. $3x + 4y < 12$

    $y > x - 2$

    $x > \text{-}2$

20. $y < 2x - 5$

    $y > \text{-}2x + 5$

    $x - 2y < 8$

CHAPTER 7

Factoring Polynomials: A Key to Success in Algebra

Rewriting a polynomial in terms of its factors is a key step in finding critical information about the polynomial. In this chapter you will learn a number of factoring rules that will work for special problems and also how to work with more generic problems. Recall that a *prime number* is an integer greater than 1 whose only factors are 1 and itself. A *prime polynomial* is one that cannot be rewritten with factors whose degree is less than that of the original polynomial.

Common Factors

The distributive property of multiplication over addition tells you how to expand a product, and it also shows you how to rewrite a problem in terms of its factors. You have learned that the statement $3(4x-5)$ can be written as $12x-15$. It is also the case that the binomial $12x-15$ can be written as the product $3(4x-5)$. Here 3 is the common factor of $12x$ and 15 and thus is "factored out." You should always remove the greatest common factor when removing a common factor.

RULE

The first rule of factoring is *always* to look for common factors.

The common factor may be a constant, a monomial, or a polynomial, but you will not often encounter a common factor that has more than a second-degree (quadratic) factor.

Example: Factor $12x^3-8x$.

$4x$ is the greatest common factor. Therefore,
$12x^3-8x=4x(3x^2-2)$.

Factor: $14x^4-21x^3+35x^2$

The greatest common factor of 14, 21, and 35 is 7. The greatest common factor of x^4, x^3, and x^2 is x^2. Therefore, the greatest common factor is $7x^2$. Accordingly, $14x^4-21x^3+35x^2=7x^2(2x^2-3x+5)$.

Factor: $12x(3x-2)+9(3x-2)$

$3x-2$ is common to both terms in this problem and is a common factor. Also, 3 is a common factor for $12x$ and 9. This makes the greatest common factor of the problem $3(3x-2)$, and $12x(3x-2)+9(3x-2)=3(3x-2)(4x+3)$.

Because $5x+7$ has no common factors, it is an example of a prime polynomial.

Difference of Squares

The product formula $(a+b)(a-b)=a^2-b^2$ is also the factoring formula $a^2-b^2=(a+b)(a-b)$. The difference of squares is easily identified because, as the name indicates, both numbers are squares and they are subtracted from each other. x^2-9 is a difference of squares but x^2+9 and x^2-19 are not. One is the sum of squares, and the other contains 19, which is not the square of an integer.

Factor: x^2-9

$$x^2-9=(x)^2-(3)^2=(x+3)(x-3)$$

Factor: $121y^2-324z^4$

$$121y^2-324z^4=(11y)^2-(18z^2)^2=(11y+18z^2)(11y-18z^2)$$

Factor: $75x^2-243$

At first glance, you note that 75 is not a square. However, when you look more closely, you realize that 75 and 243 have a common factor of 3. Factoring out 3, you arrive at the following solution:

$$75x^2-243=3(25x^2-81)=3((5x)^2-(9)^2)=3(5x+9)(5x-9)$$

Factor: $(3x-5)^2-49$

$(3x-5)^2-49$ is the difference of squares, so $(3x-5)^2-49=(3x-5)^2-(7)^2=(3x-5+7)(3x-5-7)=(3x+2)(3x-12)$.

Is this the final answer? Note that 3 is a common factor in $3x-12$, so when this 3 is factored out, the final answer becomes $3(3x+2)(x-4)$. (The monomial 3 is listed first so that it is easier to read and is not omitted from the response.)

Square Trinomials

The multiplication formulas $(a+b)^2=a^2+2ab+b^2$ and $(a-b)^2=a^2-2ab+b^2$ are called squared trinomials because the left-hand side of the equation is a square and there are three terms on the right-hand side of the equation. Whenever you see a trinomial whose first and last terms are squares and whose middle term is twice the product of the square roots of the first and last terms, the problem can be factored according to these formulas.

RULE

The sign of the middle term in the trinomial matches the sign between the terms in the binomial.

Factor: $16x^2+24x+9$

The first and last terms are squares, so you can rewrite $16x^2+24x+9$ as $(4x)^2+2(4x)(3)+3^2$. You now need to check whether the middle terms match: Is $24x$ equal to $2(4x)(3)$? Yes, it is. You now know that $16x^2+24x+9=(4x+3)^2$.

Factor: $16x^2+22x+9$

The first and last terms are squares, so you can rewrite $16x^2+22x+9$ as $(4x)^2+2(4x)(3)+3^2$. You now need to check whether the middle terms match: Is $22x$ equal to $2(4x)(3)$? No, it is not, so this trinomial is not a perfect square. Does the trinomial factor? Maybe it does (and maybe it doesn't), but that will be examined shortly. For now, you know only that the trinomial does not factor according to the square trinomial formula.

Factor: $45z^2-60zy+20y^2$

Clearly, the first and last numbers in this example are not squares. However, you did notice that the three terms are all multiples of 5, so you removed the common factor to get

$45z^2 - 60zy + 20y^2 = 5(9z^2 - 12zy + 4y^2) = 5((3z)^2 - 2(3)(2)zy + (2y)^2)$. Because $12 = 2(3)(2)$, the middle terms match.
$45z^2 - 60zy + 20y^2 = 5(3z - 2y)^2$

Quadratic Trinomials with $a = 1$

Factoring trinomials that do not fit the special pattern of the square trinomials requires that you exercise a little more care. Before you attempt to factor, stop and take a look at the pattern you get when you multiply $(x+p)(x+q)$.

$(x+p)(x+q) = x(x+q) + p(x+q)$
$= x^2 + qx + px + pq = x^2 + (p+q)x + pq$

The middle term, $p+q$, is the sum of the constants in the two factors, and the last term, pq, is the product of the two constants.

Factor: $x^2 + 11x + 30$

The two numbers that add to 11 and multiply to 30 are 5 and 6. Therefore, $x^2 + 11x + 30 = (x+5)(x+6)$.

Factor: $x^2 - x - 30$

The two numbers that add to -1 and multiply to -30 are 5 and -6. Therefore, $x^2 - x - 30 = (x+5)(x-6)$.

Factor: $x^2 - 20x + 75$

Finding the numbers that add up to -20 and multiply to 75 might take a little more thinking. You know that both numbers have to be negative (the product is positive—same signs—and the sum is negative). Listing the factors of 75, which are 1, 75; 3, 25; and 5, 15, enables you to discover that 5 and 15 are the appropriate numbers.

$$x^2 - 20x + 75 = (x - 5)(x - 15)$$

Factor: $5x^2 + 65x + 210$

Take out the common factor of 5 to get $5x^2 + 65x + 210 = 5(x^2 + 13x + 42)$. 6 and 7 are the numbers that add up to 13 and multiply to 42, so $5x^2 + 65x + 210 = 5(x + 6)(x + 7)$.

Quadratic Trinomials with $a \neq 1$

The factoring problems can get much more interesting when the quadratic coefficient is not 1 and is not a common factor of the trinomial. There are a few techniques that work to factor quadratics of the form $ax^2 + bx + c$, and in this section we will examine two of them.

Multiply: $(mx + p)(nx + q)$

$$= (mx + p)(nx) + (mx + p)(q)$$

$$= mnx^2 + pnx + qmx + pq$$

$$= mnx^2 + (pn + qm)x + pq$$

Trial and Error

In this approach, we write the factors of the coefficient a (which are m and n) and the factors of the coefficient c (which are p and q) and try to match them up to get the coefficient b (which is $pn + qm$).

Factor: $12x^2 + 23x + 10$

The factors of 12 are 1, 12; 2, 6; and 3, 4. The factors of 10 are 1, 10 and 2, 5. One combination of these factors will yield 23, or the polynomial cannot be factored. It will help if you pay attention to some old arithmetic knowledge. The sum of two even numbers is even, and the sum of two odd numbers is even. The middle term in this problem (23) is odd, so one of the results pn and qm must be even, and one must be odd. This enables you to determine that the factors 2 and 6 that give 12 are not part of the answer. Because 12 times any number or 10 times any number gives a big answer, you may want to consider those later. That is to say, the pair 3 and 4 and the pair 2 and 5 seem to be better choices than any other pair. Because $4 \times 2 = 8$, $3 \times 5 = 15$, and $8 + 15 = 23$, you can say that $m = 4$, $n = 3$, $q = 4$, and $p = 5$. Therefore,

$$12x^2 + 23x + 10 = (4x + 5)(3x + 4)$$

Factor: $4x^2 - 8x - 21$

The factors of 4 are 1, 4 and 2, 2; the factors of 21 are 1, 21 and 3, 7. The middle term is even, so the two numbers added together must be even. This eliminates 1, 4 as factors of 4, because all the factors of 21 are odd. You would get one odd and one even pair for the numbers mp and nq. Therefore, you know that $m = 2$ and $n = 2$. You also know that mp and nq are opposite in sign, because $pq = -21$. Since 21 would lead to a large result and the middle term is only -8, the reasonable factors of 21 to choose are 3 and 7. Finally, $2 \times 3 = 6$ and $2 \times (-7) = -14$ and $6 + (-14) = -8$, so you get the result:

$$4x^2 - 8x - 21 = (2x - 7)(2x + 3)$$

Splitting the Middle

The second technique in this section attempts to remove some of the guesswork that goes along with trial and error. In the process of splitting the middle, you multiply the quadratic coefficient, a, by the constant, c. You then try to find factors of this number that add up to b. Consider the following example.

Factor: $4x^2 - 8x - 21$

The product of the quadratic coefficient, 4, and the constant, -21, is -84. What are the factors of -84 whose sum is -8? Ignoring the signs for the moment, the factors of 84 are 1, 84; 2, 42; 3, 28; 4, 21; and 6, 14. Because $6 - 14 = -8$, the two numbers you are looking for are 6 and -14.

Directions	Action
Rewrite:	$4x^2 - 8x - 21 = 4x^2 - 14x + 6x - 21$
Group the first two terms and the last two terms.	$(4x^2 - 14x) + (6x - 21)$
Find the common factor for each group.	$2x(2x - 7) + 3(2x - 7)$
Remove the common binomial factor.	$(2x - 7)(2x + 3)$

You need to be careful when you split the middle. Note that if you had written $4x^2 - 8x - 21$ as $4x^2 + 6x - 14x - 21$ and then grouped, the second group would have introduced a set of parentheses after a negative. Either factor out the negative to get $(4x^2 + 6x) - (14x + 21)$ or leave the negative inside the parentheses to get $(4x^2 + 6x) + (-14x - 21)$. In this last case, the common factor for the second group is -7, not 7.

Factor: $12x^2 + 23x + 10$

The product $ac = 120$. The factors of 120 are 1, 120; 2, 60; 3, 40; 4, 30; 5, 24; 6, 20; 8, 15; and 10, 12. The pair you need is 8, 15. Split and group the middle to get

$(12x^2 + 8x) + (15x + 10)$

Common factors: $4x(2x + 3) + 5(2x + 3)$

Common factor: $(2x + 3)(4x + 5)$

Using Technology to Factor

Consider the following problem. Factor: $48x^2 + 110x + 63$. Using guess and check, you would write all the factors of 48 and 63 and then examine them to find the correct pair to match up. If you discover that this is something you do well and can do quickly, it is the easiest approach to factoring.

However, there may often be too many different combinations of numbers for you to find the correct pair, and technology can be very useful when the numbers whose factors you are seeking are very large. The product $ac = 3024$. That is a big number and has a lot of factors. Finding the pair of factors for this number that will add to 110 will probably take a great deal of time.

Using technology to find the numbers is not too difficult. Once you find the correct set of values, splitting the middle and using common factors is straightforward.

FINDING FACTORS WITH THE TI 84

Description	Keystroke	Image
Enter 3024/x	Y=	Plot1 Plot2 Plot3 \Y1■3024/X \Y2= \Y3= \Y4= \Y5= \Y6= \Y7=
Set up table to start with 1 and count by 1	2nd WINDOW	TABLE SETUP TblStart=1 △Tbl=1 Indpnt: Auto Ask Depend: Auto Ask
Examine Table	2nd GRAPH	X | Y1 1 | 3024 2 | 1512 3 | 1008 4 | 756 5 | 604.8 6 | 504 7 | 432 X=1

The last pair of numbers on this screen is 7 and 432. These factors of 3024 add up to 439, a number that is too big. Please note that for each line on this screen, the numbers multiply to 3024 and that the sum of the numbers is getting smaller as you look down the list. Use the down arrow to reach the line in which 54 and 56 are the factors.

The square root of 3024 is approximately 54.99. Because twice this number is very close to 110, it would have been easier for you to start the table at 54 and to make the change in table values -1.

Table Setting

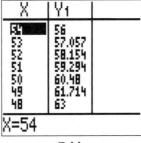

Table

With experience, you will learn to decide whether you want to begin the search from a starting value of 1 and work down or to start from the square root of the product and work up.

The search for factors with the NSpire is more easily done with the spreadsheet.

FINDING FACTORS WITH THE NSPIRE

Description	Keystrokes	Image
Open a spreadsheet. Label the column headers factor1 and Factor2.		
Enter the formula for column A in the gray row.	$=SEQ(x,x,1,\sqrt{3024})$	
Enter the formula for column B.	=3024/factor1.	
Find the sum of factor1 and factor2 in column C.	=factor1+factor2	
Scroll down column C to find the correct sum.		

Split and group the middle: $48x^2 + 110x + 63 = (48x^2 + 54x) + (56x + 63)$

Take out the common factor for each group. $6x(8x+9) + 7(8x+9)$

Take out the common factor. $(8x+9)(6x+7)$

You need to be aware of the issues when items are subtracted; you will change the formula in column C to factor1–factor2. If you are expecting an answer of -45 but get an answer of 45, you will need to adjust the signs on factor1 and factor2.

Factor: $96x^2 - 124x - 105$

The product $ac = $ -10,080. With the TI-84, Y1=-10080/x. With the NSpire, the formula for column A will need to be SEQ(x,x,1, $\sqrt{10080}$). The formula for column B on the NSpire can be factor1-factor2. Scroll down the results to get the factors 56 and -180. Split the middle, and group:

$96x^2 - 124x - 105 = (96x^2 - 180x) + (56x - 105)$

Remove the greatest common factor for each group.
$12(8x - 15) + 7(8x - 15)$

Remove the greatest common factor. $(8x - 15)(12x + 7)$

Sums and Differences of Cubes

The product formulas $(a+b)(a^2 - ab + b^2) = a^3 + b^3$ and $(a-b)(a^2 + ab + b^2) = a^3 - b^3$ can also be used for factoring. You are probably more familiar with the squares of integers than with the cubes of integers, but it is worthwhile to know the cubes of the first ten counting numbers. Like problems involving the difference of squares, these problems are easy to spot because there are only two terms. When writing the result, however, you must be careful not to confuse the trinomial in these factors with the trinomials in the square trinomial factors.

RULE

The sign in the linear binomial matches the sign between the cubes and is opposite the sign of the middle term in the trinomial.

Factor: $125x^3 + 64y^3$

$125x^3 + 64y^3 = (5x)^3 + (4y)^3$
$= (5x + 4y)((5x)^2 - (5x)(4y) + (4y)^2)$
$= (5x + 4y)(25x^2 - 20xy + 16y^2)$

Factor: $81y^6 - 3000y^3$

You know that 81 is not a cube, but you note that 81 and 3000 have a common factor of 3. You can also see that y^3 is common to both y^3 and y^6. Removing the common factor yields

$81y^6 - 3000y^3 = 3y^3(27y^3 - 1000) = 3y^3 \, (3y^2)^3 - (10)^3$
$= 3y^3 \, (3y^2 - 10)((3y)^2 + (3)(10y) + (10^2))$
$= 3y^3 \, (3y^2 - 10)(9y^2 + 30y + 100)$

Completely Factoring Polynomials

It is not always the case that a problem is completely factored after you have completed one step. Consider the following example:

Factor completely: $12x^3 + 33x^2 - 9x$

Removing the common factor of $3x$ yields $3x(4x^2 + 11x - 3)$.

However, the quadratic $4x^2 + 11x - 3 = (4x - 1)(x + 3)$.

Therefore, $12x^3 + 33x^2 - 9x = 3x(4x - 1)(x + 3)$.

Factor: $4w^4 + 11w^2 - 3$

Your first look at this problem may make you think it is very different from all the previous problems you have done. However, a second observation will show you that there are three terms. The variable in the first term is the square of the variable in the middle term, and the last term is a constant. In fact, this problem looks very much like the factor $4x^2 + 11x - 3$ in the last problem. Applying the same logic to this problem as you did to the last problem and replacing x with w^2, you get:

$$4w^4 + 11w^2 - 3 = (4w^2 - 1)(w^2 + 3)$$

RULE

Problems such as $ax^{2n} + bx^n + c$ are called quadratic in form because they look like $ax^2 + bx + c$.

As much as that is to take in (and it *is* a major transition), you have not finished solving the problem. $4w^2 - 1$ is the difference of two squares and can be factored further.

$$4w^4 + 11w^2 - 3 = (4w^2 - 1)(w^2 + 3) = (2w - 1)(2w + 1)(w^2 + 3)$$

Completely factor: $3x^3 + 5x^2 - 12x - 20$

There are no common factors for all the terms, and there are more than three terms, so you might want to say that this problem cannot be done. There are four terms, though, and you reflect that splitting the middle does work with four terms, so the idea of grouping the first two terms and the last two terms might work.

$$(3x^3 + 5x^2) - (12x + 20) = x^2(3x + 5) - 4(3x + 5) = (3x + 5)(x^2 - 4)$$

You recognize that $x^2 - 4$ is the difference of squares. The final answer to this problem is

$$3x^3 + 5x^2 - 12x - 20 = (3x - 5)(x - 2)(x + 2)$$

Exercises for Chapter 7

Completely factor each of the following:

1. $8x^2 - 24x$

2. $8x^2 - 12x$

3. $8x^4 - 12x^2$

4. $x^2 - 121$

5. $81x^2 - 25$

6. $81x^2 - 225$

7. $16p^2 - 40p + 25$

8. $64r^2 + 112r + 49$

9. $18m^3 + 60m^2 + 50m$

10. $125t^3 + 64$

11. $27 - 8q^3$

12. $729x^6 - 64$

13. $x^2 - 12x + 35$

14. $g^2 + 19g + 60$

15. $16c^2 - 62c - 45$

16. $36x^2 + 101x + 45$

17. $20q^2 - 41q + 20$

18. $20q^2 - 9q - 20$

19. $2m^3 - 2m^2 - 70m$

20. $50x^2 - 5x - 36$

21. $20x^2 - 29x - 36$

22. $x^3 - 5x^2 + 3x - 15$

23. $5x^3 + 20x^2 - 3x - 12$

24. $9x^3 + 45x^2 - 4x - 20$

25. $288n^3 + 100n^2 - 252n$

CHAPTER 8

Solving Quadratic Equations

Equations involving polynomials of degree 2 or higher are more difficult to solve than linear equations. A critical, but very simple, rule is the key to solving these equations. This is the zero product property. The ***zero product property*** means that if the product of two or more numbers is zero, then at least one of the numbers must be zero. You will learn to solve quadratic equations by factoring and using the zero product property. You will also learn how to solve quadratic equations that cannot be factored.

The Zero Product Property

You play a game with a friend. You tell her, "I am thinking of two numbers and the product of these numbers is 10. What are my numbers?" Your friend answers, "2 and 5." "No," you reply.

"1 and 10?"

"No."

"-1 and -10?"

"No."

Being a nice person, you stop the game and tell your friend, "I really wasn't thinking of any numbers. I was just going to say no to whatever you said, because there are an infinite number of pairs of numbers that multiply to yield a product of 10. For example, 1/2 times 20 is 10, and 1/10 times 100 is 10."

Do you understand the game? Let's change the game in one way by saying that the product of the two numbers is 0. Now what do you know about these two numbers?

You can be absolutely certain that at least one of the numbers is 0, because the only way for the product of two numbers to be 0 is for at least one of those two numbers to be 0. That is, if $ab=0$, then $a=0$ or $b=0$ (or both are 0).

Solve: $(x-1)(x+6)=0$

By the zero product property, either $x-1=0$ or $x+6=0$. Therefore, either $x=1$ or $x=$ -6.

Solve: $(2x+3)(4x-5)=0$

Either $2x+3=0$ or $4x-5=0$.

Either $2x=$ -3 or $4x=5$.

Either $x=$ -3/2 or $x=5/4$.

This is usually written as $x=$ -3/2, 5/4 with the word "or" implied as you read the answer.

When solving polynomial equations, set one side equal to 0, factor the polynomial, set each factor equal to 0, and solve.

Solve: $4x^2 = 9$

$4x^2 - 9 = 0$

$(2x + 3)(2x - 3) = 0$

$2x + 3 = 0$ or $2x - 3 = 0$

$2x = -3$ or $2x = 3$

$x = -3/2,\ 3/2$

Solve: $x^2 - 20x + 75 = 0$

$(x - 5)(x - 15) = 0$

$x - 5 = 0$ or $x - 15 = 0$

$x = 5,\ 15$

Solve: $12x^3 + 33x^2 - 9x = 0$

$3x(4x^2 + 11x - 3) = 0$

$3x(4x - 1)(x + 3) = 0$

$3x = 0$ or $4x - 1 = 0$ or $x + 3 = 0$

$x = 0,\ 1/4,\ -3$

Completing the Square with $a = 1$

When asked to solve an equation such as $4x^2 = 9$, you may have thought about dividing both sides of the equation by 4 to get $x^2 = 9/4$. Then you may have reasoned that all you needed to do was take the square root of both sides to get the answer $x = 3/2$. This is not correct. Your equation $x^2 = 9/4$

asks for those numbers that give the answer 9/4 when squared, and -3/2 also solves this problem!

ESSENTIAL

When taking the square root of both sides of an equation, you must remember that there are both a positive number and a negative number that will solve the equation.

Solve: $x^2 = 5$

There are two numbers that solve this problem, $\sqrt{5}$ and $-\sqrt{5}$.

What does one do if the quadratic in three terms does not factor? For example, suppose the problem is $x^2 + 6x + 3 = 0$. There are no integers that multiply to yield the product 3 and add up to 6. The technique for solving this problem is called ***completing the square*** trinomial. That is, you are going to make the left-hand side of the equation fit the pattern of the square trinomial $a^2 + 2ab + b^2$. Observe that the first two terms in $x^2 + 6x + 3$ are similar to the first two terms in $a^2 + 2ab + b^2$. If $a = x$, then $2b$ must equal 6, so $b = 3$ and $b^2 = 9$.

Here is how to solve $x^2 + 6x + 3 = 0$ by this process:

COMPLETE THE SQUARE WITH $a = 1$

Description	Equation
Move the constant to the other side of the equation.	$x^2 + 6x + = -3$
Add b^2 to both sides of the equation.	$x^2 + 6x + 9 = -3 + 9$
Factor the left side, and simplify the right side.	$(x+3)^2 = 6$
Take the square root of both sides of the equation.	$x + 3 = \pm\sqrt{6}$
Solve for x.	$x = -3 \pm\sqrt{6}$

Solve by completing the square: $x^2 - 8x + 9 = 0$

Move the constant. $x^2 - 8x = \text{-}9$

Take half the linear coefficient (b),
square it, and add it to both sides. $x^2 - 8x + 16 = \text{-}9 + 16$

Factor and simplify. $(x-4)^2 = 7$

Take the square root. $x - 4 = \pm\sqrt{7}$

Solve for x. $x = 4 \pm\sqrt{7}$

The trouble arises when the linear coefficient is odd. When this happens, leave half the linear term as a fraction.

Solve by completing the square: $x^2 - 7x - 6 = 0$

Move the constant. $x^2 - 7x = 6$

Take half the linear coefficient (b),
square it, and add it to both sides. $x^2 - 7x + \left(\dfrac{7}{2}\right)^2 = 6 + \left(\dfrac{7}{2}\right)^2$

Factor and simplify. $\left(x - \dfrac{7}{2}\right)^2 = 6 + \dfrac{49}{4} = \dfrac{73}{4}$

Take the square root of both sides. $x - \dfrac{7}{2} = \pm\sqrt{\dfrac{73}{4}} = \pm\dfrac{\sqrt{73}}{2}$

Solve for x. $x = \dfrac{7}{2} \pm \dfrac{\sqrt{73}}{2} = \dfrac{7 \pm \sqrt{73}}{2}$

Completing the Square with *a* ≠ 1

If you are confronted by a case where the quadratic coefficient is not 1, the first step is to divide both sides of the equation by the quadratic coefficient. Then continue to complete the square as you have just practiced.

Solve: $4x^2 + 32x - 27 = 0$

Divide by 4.	$x^2 + 8x - 27/4 = 0$

Move the constant.
$$x^2 + 8x = 27/4$$

Take half the linear coefficient (b),
square it, and add it to both sides.
$$x^2 + 8x + 16 = 27/4 + 16$$

Factor and simplify.
$$(x+4)^2 = 91/4$$

Take the square root of both sides.
$$x + 4 = \pm\sqrt{\frac{91}{4}}$$

Solve $x = -4 \pm\sqrt{\dfrac{91}{4}} = -4 \pm\sqrt{\dfrac{91}{2}}$

Solve by completing the square.
$$12x^2 + 23x + 10 = 0$$

Divide by 12.
$$x^2 + \frac{23}{12}x + \frac{10}{12} = 0$$

Move the constant, take half the linear coefficient (b), square it, and add it to both sides:

$$x^2 + \frac{23}{12}x + \left(\frac{23}{24}\right)^2 = \frac{-10}{12} + \left(\frac{23}{24}\right)^2$$ Do you see that half of 23/12 is 23/24?

Factor and simplify.
$$\left(x + \frac{23}{24}\right)^2 = \frac{-10}{12} + \frac{529}{576} = \frac{49}{576}$$

Take the square root.
$$x + \frac{23}{24} = \pm\frac{7}{24}$$

Solve. $x = 30/24 = 5/4$ or $x = 16/24 = 2/3$

Because the solutions to the problem (or the ***roots of the equation***) are rational numbers, this problem could have been solved by factoring and using the zero product property.

The Quadratic Formula

Because the steps in solving a quadratic equation by completing the square always involve the same pattern, you can create a formula for solving all quadratics of the form $ax^2 + bx + c = 0$. This equation is called the **quadratic formula**.

Solve $ax^2 + bx + c = 0$.

$$x = \frac{\text{-}b \pm \sqrt{b^2 - 4ac}}{2a}$$

ALERT

Remember to put parentheses around the b when computing b^2, especially when using technology.

Solve using the quadratic formula: $4x^2 - 8x - 21 = 0$

$a = 4$, $b = \text{-}8$, and $c = \text{-}21$

$$x = \frac{-(\text{-}8) \pm \sqrt{(\text{-}8)^2 - 4(4)(\text{-}21)}}{2(4)}$$

$$x = \frac{8 \pm \sqrt{400}}{8}$$

$$x = \frac{8 \pm 20}{8} = \frac{8 + 20}{4}, \frac{8 - 20}{4} = \frac{7}{2}, \frac{\text{-}3}{2}$$

Solve using the quadratic formula: $5x^2 - 18x - 20 = 0$

$$x = \frac{-(\text{-}18) \pm \sqrt{(\text{-}18)^2 - 4(5)(\text{-}20)}}{2(5)}$$

$$x = \frac{18 \pm \sqrt{724}}{10}$$

Exercises for Chapter 8

Solve.

1. $x(x-4)=0$

2. $(3x+2)(5x-3)=0$

3. $(4x+1)(5x-6)(9x-4)=0$

Solve using the zero product property.

4. $3x^2-5x-2=0$

5. $3x^2-5x+2=0$

6. $30x^2+7x-2=0$

7. $45x^2+9x-8=0$

Solve by completing the square.

8. $x^2+6x-5=0$

9. $x^2+7x-5=0$

10. $12x^2+24x-15=0$

Solve by using the quadratic formula.

11. $12x^2+17x-40=0$

12. $12x^2+17x-49=0$

13. $6x^2-34x+31=0$

14. $15x^2-29x+13=0$

Solve by any method you choose.

15. $8x^2 = 72$

16. $9x^2 = 72x$

17. $15x^2 - 29x + 12 = 0$

18. $18x^2 + 27x - 35 = 0$

19. $42x^2 + 19x = 100$

CHAPTER 9

Quadratic Relationships

Quadratic functions have a number of applications. When Isaac Newton saw the apple fall in his orchard, he realized that an item in free fall could be used to measure gravity. His experiments showed that the height of an object in free fall, the path of a cannon ball, or the path of a ball thrown by a person can be modeled with quadratic functions. Police can use the length of skid marks on a road and the weather conditions to determine the speed at which a car was traveling when the brakes were applied.

The Parabola in Standard Form $y = ax^2 + bx + c$

Enter the equation $y = x^2$ into your calculator and graph this function.

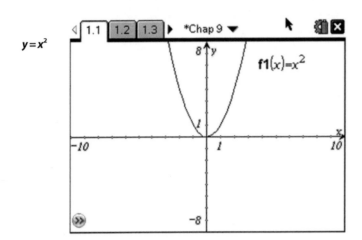

y=x²

The graph of this function is called a parabola. The turning point, or
vertex, for the parabola is at the origin. Rewrite the equation with a coefficient for x^2. Try numbers such as 2, 3, 1/2, 0.1, -1, and -1/2. How would
you compare the graphs of $y = 2x^2$ and $y = 3x^2$ to the parent graph $y = x^2$?
You might respond that the graphs are narrower or rise more steeply. The
graphs of $y = 1/2x^2$ and $y = 0.1x^2$ might be described as wider or rising at
a slower rate. The graphs of $y = -1x^2$ and $y = -1/2x^2$ are first described as
being flipped over the x-axis, and then a statement about width is considered. The important thing is that the quadratic coefficient changes the rate
at which the parabola grows.

Rewrite the function as $y = x^2 + 1$ and graph it. The basic parabola is
translated vertically 1 unit up. Change the constant from 1 to 2, to 3, to -1,
and to -2. The graph of the basic function is translated up when the constant is positive and down when the constant is negative. The y-intercept
of this graph is the constant. The constant at the end of the parabola causes
a vertical shift in the parabola.

Rewrite the function as $y = x^2 - 2x$ and graph it. The vertex of the
parabola is no longer at the origin, nor is it on the y-axis. Change the coefficient of x to -4, -5, -6, 2, 3, and 4. (That is, the equation will be $y = x^2 - 4x$,
$y = x^2 - 5x$, $y = x^2 - 6x$, $y = x^2 + 2x$, $y = x^2 + 3x$, and $y = x^2 + 4x$.) The graph

slides to the right when the linear coefficient is negative and to the left when it is negative.

You now have some idea of the impact that each of the coefficients in $y=ax^2+bx+c$ has on the parent function $y=x^2$.

RULE

> You observed that when the value of the quadratic coefficient, *a*, is positive in the graph of $y=ax^2$, the parabola opens upward and that when *a* is negative, the graph opens downward. The term used to describe this curvature is called **concavity**. When *a* is positive, the graph of $y=ax^2$ is concave up, and when *a* is negative, the graph is concave down.

Finding the Axis of Symmetry and the Vertex

Enter the equation for $y=x^2$ into your calculator, and look at a table of values.

$y=x^2$ with
Table of Values

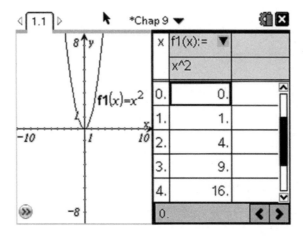

You should not be surprised to observe that when the values of x are negatives of one another, the y values are the same. The impact of this simple fact is that the vertical line passing through the vertex serves as a line of symmetry for the parabola. Take a piece of graph paper, plot these points, and graph the parabola. Fold the paper along the y-axis, and observe that when you do so, the two arcs of the parabola lie one on top of the other. The vertex of the parabola is the only point on the graph that does not have a mirror image.

Where is the axis of symmetry when the graph is moved vertically? It is still the y-axis. Where is the axis of symmetry when the graph is moved horizontally? It makes sense that the axis of symmetry also moves horizontally, but finding its equation can save you a lot of work. Take a look at the table of values for the function $y=x^2-2x$. For what value of x do you see the same type of symmetry you saw with $y=x^2$?

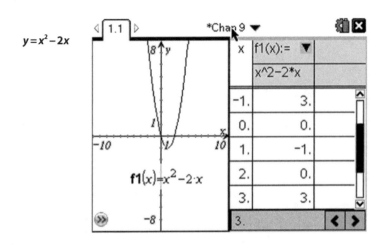

The y values are symmetric about $x=1$. Repeat this process for $y=x^2-4x$, $y=x^2+2x$, and $y=x^2+4x$. The axes of symmetry are at $x=2$, $x=-1$, and $x=-2$, respectively. Does this help you see that when the equation of the parabola is $y=ax^2+bx+c$, the equation for the axis of symmetry will be $x=-b/2a$?

ALERT

Writing the *equation* for the axis of symmetry is often a place where people lose points on tests, because they fail to write $x=$ as part of their response.

Given the equation $y = x^2 - 3x$, the formula states that the axis of symmetry should have the equation $x = 3/2$. Check to see that this agrees with your graph and its table of values.

Determine the equation for the axis of symmetry for the graph of $y = 3x^2 + 12x - 1$.

$a = 3$, $b = 12$. The equation for the axis of symmetry is $x = -12/2(3)$, or $x = -2$.

The vertex of the parabola is the point where the axis of symmetry intersects the parabola. Substitute the value for the axis of symmetry into the equation to find the coordinates of the vertex.

Determine the coordinates of the vertex of the parabola with equation $y = 3x^2 + 12x - 1$.

The axis of symmetry is at $x = -2$. The vertex is at $y = 3(-2)^2 + 12(-2) - 1 = -13$. The coordinates of the vertex are (-2, -13).

Find the equation for the axis of symmetry and the coordinates of the vertex of the parabola with equation $y = -1/3 \, x^2 + 2x + 2$.

$a = -1/3$ and $b = 2$. The equation for the axis of symmetry is $x = -2/2$ $(-1/3) = 3$. The y-coordinate for the vertex is $y = -1/3(3)^2 + 2(3) + 2 = 5$. The coordinates of the vertex are (3, 5).

The Transformed Parabola: Vertex Form $y = a(x - h)^2 + k$

Enter the equation $y = (x - 1)^2 - 2$ into your calculator and sketch it. What is the equation for the axis of symmetry? What are the coordinates of the vertex? The axis is at $x = 1$, and the vertex is at the point (1, -2). The vertex form for the parabola is $y = a(x - h)^2 + k$. This equation tells you that the axis of symmetry has equation $x = h$ and that the coordinates of the vertex are (h, k). To help you remember this, the axis of symmetry for the original parabola was $x = 0$. Set the expression inside the parentheses equal to zero and solve. Because this value is also the x-coordinate for the vertex, when you substitute it into the equation, the number inside the parentheses will be zero; multiplied by a it will still be zero, and when k is added the result is k.

Find the equation for the axis of symmetry and the coordinates of the vertex for $y=2(x+3)^2+2$.

Axis: $x+3=0$ yields $x=$-3.

Vertex: $y=$-$2($-$3+3)^2+2=2$. The coordinates of the vertex are (-3, 2).

The process of completing the square is used to convert equations of the form $y=ax^2+bx+c$ to vertex form. Be aware that, in contrast to the form we use when solving equations, one side of this equation is not zero but is the variable y instead.

REWRITE IN VERTEX FORM: $y=3x^2+12x-4$

Description	Action
Divide by the quadratic coefficient.	$\dfrac{1}{3}y=x^2+4x-\dfrac{4}{3}$
Move the constant to the other side.	$\dfrac{1}{3}y+\dfrac{4}{3}=x^2+4x$
Add half the linear coefficient to both sides.	$\dfrac{1}{3}y+\dfrac{4}{3}+4=x^2+4x+4$
Factor and simplify.	$\dfrac{1}{3}y+\dfrac{16}{3}=(x+2)^2$
Solve for y.	$\dfrac{1}{3}(y+16)=(x+2)^2$
	$y+16=3(x+2)^2$
	$y=3(x+2)^2-16$

The axis of symmetry is $x=$-2, and the coordinates of the vertex are (-2, -16).

Factoring by Graphing

You can use the graphs of functions to help with factoring them. Consider the following problem:

Solve: $4x^2 + 9x + 2 = 0$

Factor. $(4x + 1)(x + 2) = 0$

Use the zero product property. $4x + 1 = 0$, or $x + 2 = 0$

Solve. $x = -1/4, -2$

Graph the equation $y = 4x^2 + 9x + 2$.

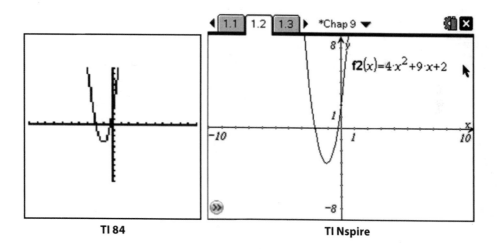

TI 84   TI Nspire

Finding the **zeroes** of the function is the same as finding the **x-intercepts** for the graph. Use the Zero feature with the TI 84 and the Points of Intersection feature with the TI Nspire.

FINDING ZEROES

TI 84	TI Nspire
Press 2nd TRACE to go to the CALC menu.	Press Menu
Choose option 2: Zero.	Choose Option 7: Points & Lines
When prompted for the left bound, enter -5.	Choose Option 3: Intersection Point(s)
When prompted for the right bound, enter -1.	Move the cursor to the parabola, enter
Press ENTER for guess.	Move the cursor to the x-axis, enter
The zero is -2. (Note that $y = 1E\text{-}12$ is the number 1×10^{-12}. This should be zero, but there are round-off errors within the calculator that you need to accept.)	ESC to leave this menu.
Repeat the process with the TI 84 to find the second root. Use -1 for the left bound and zero for the right bound.	Press Menu, Option 1: Actions, Option 7: Coordinates and Equations.
	Move the mouse to the first point of intersection, click the mouse, move the cursor to a convenient place on the screen to display the coordinates, and click.
	Repeat for the second point of intersection.

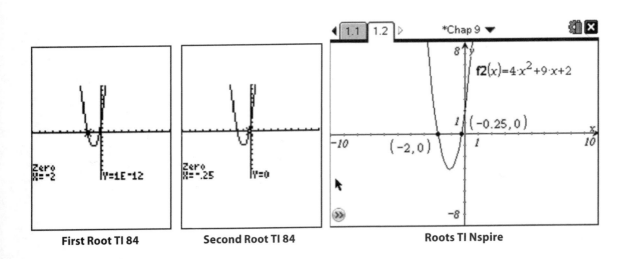

First Root TI 84	Second Root TI 84	Roots TI Nspire

ALERT

Your calculator can convert decimals to fractions. Consult your calculator manual to learn how to convert a decimal to a fraction.

The decimal -0.25 is -1/4. The zeroes of the function are:

$x = $ -2, -1/4

Rewrite as two equations.	$x = $ -2 or $x = $ -1/4
Rewrite.	$x + 2 = 0$ or $4x = $ -1, which becomes $4x + 1 = 0$
Write as a product.	$(x + 2)(4x + 1) = 0$

Answer the factoring question. $4x^2 + 9x + 2 = (x + 2)(4x + 1)$

Circles

The equation of a circle with center (h, k) and radius r is $(x - h)^2 + (y - k)^2 = r^2$. The process of completing the square will enable you to rewrite the standard form of the equation of the circle $Ax^2 + Ay^2 + Bx + Cy + D = 0$ in the center-radius form.

Find the coordinates of the center and the length of the radius of the circle with equation $x^2 + y^2 - 8x + 10y - 23 = 0$.

Gather the terms in x together, gather the terms in y together, and move the constant to the right-hand side of the equation.

$$x^2 - 8x + y^2 + 10y = 23$$

Complete the square in each variable.

$$x^2 - 8x + \mathbf{(4)^2} + y^2 + 10y + \mathbf{(5)^2} = 23 + \mathbf{(4)^2} + \mathbf{(5)^2}$$

$$(x - 4)^2 + (y + 5)^2 = 64$$

The center of the circle is (4, -5), and the radius has length 8.

Extreme Applications of the Quadratic Function

The height of an object thrown vertically upward with an initial velocity of 96 feet per second from a height of 50 feet above the ground can be represented by the formula $h(t) = -16t^2 + 96t + 50$, where t is the time, in seconds, that the object is in the air. You can see the 96 ft/sec and 50 feet in the equation. The quadratic coefficient -16 represents the gravitational effect that causes the ball to return to the ground.

What is the maximum height of the object?

You can determine the answer to this problem by entering the equation into your calculator, graphing the function, and then determining the maximum value of y in the graph. You can also solve this problem by realizing that the graph is a parabola that is concave down and that the vertex of the parabola will be the answer.

Axis of symmetry: $t = -96/2(-16) = 3$. The object reaches maximum height in 3 seconds. Vertex: $h(3) = -16(3)^2 + 96(3) + 50 = 194$ feet.

What is the height of the object after 2 seconds? What is the height of the object after 4 seconds?

$h(2) = -16(2)^2 + 96(2) + 50 = 178$ feet

$h(4) = -16(4)^2 + 96(4) + 50 = 178$ feet

Are you surprised that these heights are the same? Remember that $t = 3$ is the axis of symmetry. The point that occurs 1 second before this ($t = 2$) will correspond to the point that occurs 1 second beyond the axis ($t = 4$). The object is at a height of 178 feet on the way up at time 2 seconds *and* at a height of 178 feet on the way down at time 4 seconds.

When is the object at a height of 90 feet?

Setting $h = 90$, the equation becomes $90 = -16t^2 + 96t + 50$.

Moving terms to the left yields $16t^2 - 96t + 40 = 0$.

Use the quadratic formula:

$$t = \frac{-(-96) \pm \sqrt{(-96)^2 - 4(16)(40)}}{2(16)} = \frac{96 \pm \sqrt{6656}}{32} = 0.450, 5.550$$

The object will be at a height of 100 feet at 0.450 second and then again at 5.550 seconds.

When will the object hit the ground?

The height of the object when it strikes the ground is 0 feet. Solving $-16t^2 + 96t + 50 = 0$ yields 6.482 seconds. We ignore the negative answer to this problem, because the object is released at time $t=0$.

Something Different: $x = y^2$

Recall that when you interchange the x- and y-coordinates of a graph, the impact is to reflect the graph across the line $y=x$ and that the resulting graph represents the inverse of the original. Doing so with the parabola $y=x^2$ yields a new graph, $x=y^2$, which is a parabola opening to the right. Does this graph represent a function? No, it does not, because the graph fails the vertical-line test.

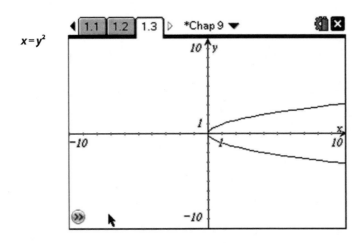

RULE

When the domain of a function is restricted, the full range of the function must be maintained.

What prevents the inverse from being a function? If only half of the original parabola were sketched, then the inverse would also be a function. Consequently, the domain of the parabola $y = x^2$ is restricted to $x \geq 0$.

$y = x^2$ and $x = y^2$ in a restricted domain

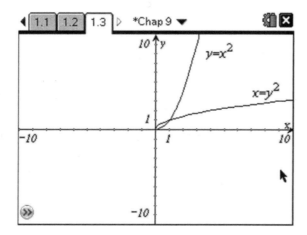

Solving the equation $x = y^2$ for y gives the very familiar equation $y = \sqrt{x}$. The square root function is the inverse of the squaring function.

Exercises for Chapter 9

Find the equation for the axis of symmetry and coordinates of the vertex for each parabola. Does the vertex represent a maximum or a minimum value for the quadratic function?

1. $y = -2x^2 + 4x + 3$

2. $y = 2x^2 + 8x - 1$

3. $y = 1/2x^2 - 4x + 3$

4. $y = -5x^2 + 24x + 31$

5. $y = -3x^2 + 8x - 2$

6. $y = -16x^2 + 112x + 94$

Exercises 7–12: Rewrite each of the equations in Exercises 1–6 in vertex form.

Exercises 13–18: Find the zeroes for each of the quadratic functions in Exercises 1–6.

19. Factor $12x^2 - x - 63$ by using the graph of the corresponding parabola.

Exercises 20–22: Determine the coordinates of the center of the circle and the length of the radius.

20. $(x+4)^2 + (y+3)^2 = 25$

21. $x^2 + y^2 - 20x + 12y + 15 = 0$

22. $2x^2 + 2y^2 + 8x - 14y - 15 = 0$

23. A ball is thrown vertically into the air with an initial velocity of 144 feet per second from a point 80 feet above ground.

   a. When does the ball reach its maximum height?

   b. What is the maximum height?

   c. When is the ball at a height of 150 feet?

   d. When is the ball at a height of 50 feet?

   e. When does the ball strike the ground?

CHAPTER 10

Quadratic Systems

The **break-even point** for a business is the point at which the business has earned exactly the amount of money it has spent to produce a product. At the break-even point, the business's profit—the difference between its revenue and its costs—is equal to 0. **Market equilibrium** is a term used in economics to describe the point at which the consumers and suppliers agree on what the price of an item will be and on how many of these items will be available for purchase. Solving systems of equations can help determine where the break-even and equilibrium points are located.

Linear-Quadratic Systems with Graphical Solutions

There are three possibilities for the point(s) of intersection when a line and a parabola or a line and a circle are graphed in the same coordinate system. They can intersect at 0, 1, or 2 points. The intersection feature of your graphing calculator can help you find the intersection of a line and a parabola quite readily. Finding the intersection of a line and a circle requires a little more work, because the circle is not a function and cannot easily be graphed on a calculator.

Solve graphically: $y = -2x^2 + 8x + 5$

$y = 10x + 1$

When you enter each of the equations into your calculator, you realize that the window dimensions on your calculator need to be changed. If you do not have a good feel for the size of the numbers the equations are producing, use the Table feature of your calculator to gauge how big your dimensions should be.

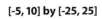

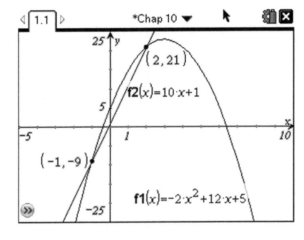

The points of intersection are (-1, -9) and (2, 21).
The circle $x^2 + y^2 = 25$ can be graphed in the function mode of the graphing calculator by solving the equation for y and entering the equations for

each of the semicircles into separate equations. That is, $y=\sqrt{25-x^2}$ is the equation for the upper semicircle, and $y=-\sqrt{25-x^2}$ is the equation for the lower semicircle.

Solve graphically: $x^2+y^2=25$

$y=x+1$

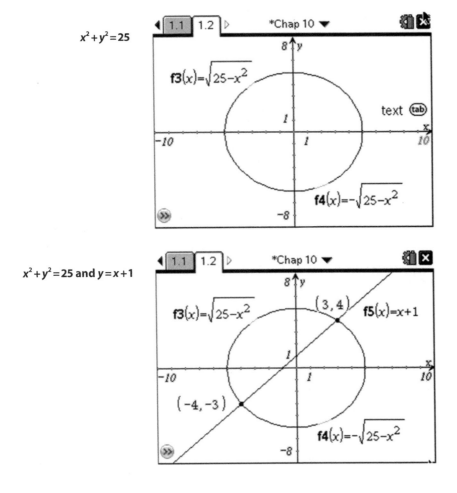

The points of intersection are (-4, -3) and (3, 4).

Linear-Quadratic Systems with Algebraic Solutions

The substitution process can be used to rewrite linear-quadratic systems algebraically into a single equation in one variable. There may be times in which fractions will be introduced into the problem, but they can be handled by multiplying both sides of the equation by a common denominator.

Solve algebraically: $y = 3x^2 + 7x - 8$

$y = 4x - 2$

ALERT

The substitution process is used more than any other method when solving a linear-quadratic system algebraically because it will quickly give a quadratic equation in one variable which can be solved.

Substituting $3x^2 + 7x - 8$ for y in the second equation gives the equation:

$3x^2 + 7x - 8 = 4x - 2$

$3x^2 + 3x - 6 = 0$

$3(x^2 + x - 2) = 0$

$3(x + 2)(x - 1) = 0$

$x = \text{-}2,\ 1$

When $x = 1$, $y = 4(1) - 2 = 2$, and when $x = \text{-}2$, $y = 4(\text{-}2) - 2 = \text{-}10$. The solution to the problem is (1, 2) and (-2, -10).

Solve algebraically: $(x + 3)^2 + (y - 2)^2 = 20$

$x - 2y = \text{-}7$

Solve for x in the linear equation to get $x=2y-7$. Substitute this for x in the equation for the circle.

$$(2y-7+3)^2+(y-2)^2=20$$

$$(2y-4)^2+(y-2)^2=20$$

$$4y^2-16y+16+y^2-4y+4=20$$

$$5y^2-20y=0$$

$$5y(y-4)=0$$

$$y=0, y=4$$

When $y=0$, $x=2(0)$ -7= -7, and when $y=4$, $x=2(4)-7=1$. The solution to the system of equations is (-7, 0) and (1, 4).

QUESTION

Why is it not wise to substitute into the quadratic equation when finding the other element of the ordered pair?
For a given of y, there may be two values of x which will satisfy the equation, one of which may not lie on the line.

Solve algebraically: $(x-2)^2+(y+3)^2=72$

$$x+2y=2$$

Solve for x in the linear equation to get $x=$ -2$y+2$. Substitute into the equation for the circle:

$$(\text{-}2y+2-2)^2+(y+3)^2=72$$

$$(\text{-}2y)^2+(y+3)^2=72$$

$$4y^2+y^2+6y+9=72$$

$$5y^2+6y-63=0$$

Solve for y by factoring as $(y-3)(5y+21)=0$ or by the quadratic formula:

$$y=\frac{-6\pm\sqrt{1296}}{10}=\frac{-6\pm36}{10}$$

$y=3, y=-4.2$

If $y=3$, then $x=-2(3)+2=-4$, and if $y=-4.2$, then $x=-2(-4.2)+2=10.4$. The solution to the system of equations is the ordered pairs (-4, 3) and (10.4, -4.2).

Applications from the Business World

Companies must balance the amount of money they spend (cost) with the amount of income they earn (revenue). Not-for-profit companies try to get their revenue and costs to be the same each year, whereas profit-making companies strive to have their revenue be larger than their cost.

For example, a business determines that its revenue function for selling x items of its product is given by the equation $R(x)=-x^2+1800x$ and that the cost of producing these x items is given by the function $C(x)=150x+350,000$, where revenue and cost are in dollars.

1. **How many items should the company sell to maximize its revenue?** This question asks for the axis of symmetry for the graph of the equation $x=-1800/-2=900$. The company needs to produce 900 items to maximize its revenue.
2. **What is the maximum revenue?** The revenue is the output of the function. $R(900)=\$810,000$
3. **How much does it cost to produce these items?** $C(900)=\$485,000$
4. **What is the profit for producing and selling these items?** The profit for producing and selling 900 items is $810,000-485,000=\$325,000$
5. **Write an equation for the profit function.**
   $$P(x)=R(x)-C(x)$$
   $$P(x)=(-x^2+1800x)-(150x+350,000)=-x^2+1650x-350,000$$
6. **At what point does the company begin to make a profit? When is the company working with a deficit (losing money)?** The com-

pany is breaking even when the profit is equal to zero (that is, when revenue $=$ cost). $P(x)=0$ when $x=250$ or $x=1400$. The company breaks even when it produces and sells 250 items. Before that point, the company is losing money. The company also breaks even when it produces and sells 1400 items. Beyond that point, the company is losing money.

7. **How many items must be produced and sold to maximize the profit?** The axis of symmetry for the profit function is $x=825$. The company will earn a maximum profit when 825 items are produced and sold.

8. **What is the maximum profit?** $P(825)=\$330{,}625$

Systems with Two Quadratic Relationships

Finding the coordinates of the points of intersection of two parabolas, of two circles, or of a circle and a parabola can also be done algebraically and graphically. The graphical approach is no different from using a linear-quadratic system of equations: Graph the functions, and use the intersection feature to find the points of intersection. The algebraic approach is also the same: Use substitution to get a single equation in one variable.

Solve the system of equations algebraically:

$$y=x^2-4x-5$$
$$y=-x^2+x+2$$

Substituting for y yields:

$$x^2-4x-5=-x^2+x+2$$
$$2x^2-5x-7=0$$
$$(2x-7)(x+1)=0$$
$$x=7/2,\ x=-1$$

When $x=3.5$, $y=-6.75$, and when $x=-1$, $y=0$. The points of intersection for these parabolas are (3.5, -6.75) and (-1, 0).

Solve the system of equations algebraically:

$$x^2 + y^2 = 20$$

$$y = 1/8\, x^2$$

Rewrite the second equation to read $8y = x^2$. Substituting into the first equation gives:

$$8y + y^2 = 20$$

$$y^2 + 8y - 20 = 0$$

$$(y + 10)(y - 2) = 0$$

$$y = \text{-}10, y = 2$$

When $y = 2$, $x^2 = 16$, so $x = \pm 4$. When $y = \text{-}10$, $x^2 = \text{-}80$. There are no real values of x for which this is true. Thus $y = \text{-}10$ is called an **extraneous** root. (In a way, this word tells you all you need to know about such a root: It is extra and it is erroneous.) The circle and the parabola intersect at the points (4, 2) and (-4, 2).

Solve the system of equations algebraically:

$$x^2 + y^2 = 25$$

$$y = 2x^2 - 2$$

Solve the second equation for x^2 to get $x^2 = \dfrac{y+2}{2}$, and substitute this into the first equation.

$$\frac{y+2}{2} + y^2 = 25$$

Multiply by 2. $\qquad\qquad\qquad$ $y + 2 + 2y^2 = 50$

Gather terms. $\qquad\qquad\qquad$ $2y^2 + y - 48 = 0$

Use the quadratic formula. $\qquad$ $y = \dfrac{\text{-}1 \pm \sqrt{1^2 - 4(2)(\text{-}48)}}{4} = \dfrac{\text{-}1 \pm \sqrt{385}}{4}$

ESSENTIAL

When using decimal approximations, never round off until you have done all the computations as the rounding may result in incorrect answers later in the problem.

These are *ugly* numbers. Trying to find the x-coordinates as radical terms will be very difficult (and also impractical). Starting with the positive solution, compute the decimal and store it in a memory location on your calculator.

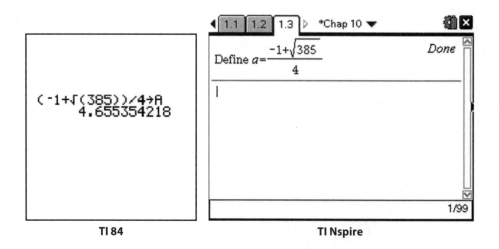

TI 84

TI Nspire

Solve for x^2 and then for x. When $y = \dfrac{-1 + \sqrt{385}}{4}$, $x^2 = 5.827677109$ and $x = \pm 2.414$, rounded to the nearest thousandth. $y = \dfrac{-1 + \sqrt{385}}{4} = 4.655$ (rounded to the nearest thousandth). Two points of intersection are (2.414, 4.655) and (-2.414, 4.655).

When $y = \dfrac{-1 - \sqrt{385}}{4}$, $x^2 = 0.9223228912$ and $x = \pm 0.960$ (rounded to the nearest thousandth). $y = \dfrac{-1 - \sqrt{385}}{4} = -5.156$ (rounded to the nearest thousandth), so the last two points of intersection are (0.960, -5.156) and (-0.960, -5.156). It would be very convenient if it were always possible to apply graphical approaches to solving systems of equations beyond the linear-quadratic, but some equations do not transfer nicely into a graphing calculator. For this reason, knowing something about algebraic solutions is essential.

Exercises for Chapter 10

Solve each of the following systems of equations graphically.

1. $y = 3x^2 - 4x - 2$

    $y = \text{-}4x + 1$

2. $y = \text{-}5x^2 + 2x + 3$

    $y = \text{-}2x + 2$

3. $y = x^2 - 9$

    $y = 16 - x^2$

Solve each of the following systems algebraically.

4. $y = x^2 - 4x - 5$

    $y = 2x - 13$

5. $y = \text{-}3x^2 + 8x + 10$

    $y = 11x + 4$

6. $y = \text{-}0.5x^2 + 2x + 3$

    $5x - 2y = \text{-}4$

7. $x^2 + y^2 = 100$

    $y = x - 2$

8. $x^2 + y^2 = 25$

   $y = x^2 - 5$

9. $x^2 + y^2 = 100$

   $4x - 3y = 0$

10. A company determines that its monthly revenue function can be expressed by the equation $R(x) = 2800x - x^2$, where x is the number of units sold.

    a. How many units must be sold to reach the maximum monthly revenue?

    b. Determine the maximum monthly revenue.
        The monthly cost function for producing x units is
        $C(x) = 2100x + 60,000$.

    c. When monthly revenue is at a maximum, how much does it cost to produce these items?

    d. What is the maximum monthly profit that the company can make for this product?

CHAPTER 11

Rational Expressions

You learned a long time ago about the need to find a common denominator when adding or subtracting fractions. You also learned that it is not necessary to find a common denominator when multiplying or dividing fractions.

Equivalent Fractions

You probably remember your teacher discussing slices of pizza as an illustration of using fractions. If your pizza is an 8-cut (that is, the pizza is cut into 8 pieces, supposedly equal in size), then each slice represented 1/8 of the pizza. If the same pizza were cut into 24 slices, how many of these slices would represent the same amount of pizza as 1 slice of an 8-cut pizza?

In order to answer the question in the previous paragraph, it is important for you to recognize the premise that the pizzas were the same size to begin with. What would happen if the pizzas were not the same size? We would have to do a lot more work to determine how many slices of each pizza result in an equal amount of food (dough, sauce, and cheese). However, no matter what the size of the pizza, we can talk about the *relative* amount of the pizza eaten. Eating 1 slice of an 8-cut pizza means that 1/8 of the pizza has been eaten. Eating 3 slices of a 24-cut pizza means that 3/24, or 1/8, of the pizza has been consumed. That is why 1/8 and 3/24 are called ***equivalent*** fractions.

Arithmetic with Fractions

Addition and subtraction involve combining common characteristics. What do you get when you add 8 apples and 3 apples? 11 apples. What is the result of taking 4 oranges from 9 oranges? 5 oranges. You can find the sum of 8 and 3 or the difference between 9 and 4 in a sterile condition. But arithmetic (and mathematics) becomes more interesting when it is "applied," or used in context.

Let's return to the example of the 24-cut pizza. If John ate 1/3 of the pizza and Sally ate 1/8 of the pizza, how much of the pizza remains? In this context, you can calculate how many slices were eaten and how many remain, and from this you can determine what fraction of the pizza remains. John ate 1/3 of the 24 slices. How many slices is this? 24 slices divided by 3 is 8 slices. Sally ate 3 slices (24 divided by 8). Together, they ate 11 slices. This leaves 13 slices, so 13/24 of the pizza remains.

From a fractional perspective, John and Sally ate $\frac{1}{3} + \frac{1}{8}$ of the pizza.

Converting these fractions into equivalent fractions with a denominator of

24, you see that they ate $\dfrac{8}{24}+\dfrac{3}{24}$ of the pizza. You can consider this to be an example of adding $8+3$ with the characteristic being twenty-fourths, rather than apples or oranges. $\dfrac{8}{24}+\dfrac{3}{24}=\dfrac{11}{24}$. How much of the pizza is left? You started with a whole pizza (1 pizza) and ate 11/24 of it. The amount left is $1-\dfrac{11}{24}$. Rewriting this with equivalent fractions yields $\dfrac{24}{24}-\dfrac{11}{24}$, so it is clear that 13/24 of the pizza is left.

If the first problem that arises in the arithmetic of fractions is the question of why you find common denominators, we hope that the pizza problem and your life experiences have given you a better understanding of why this approach is necessary. The second problem is in finding the common denominator. Here we want to emphasize that factoring will be the critical process, whether the fraction contains constants or algebraic expressions.

You recall the mechanics of finding equivalent fractions: Multiply both numerator and denominator by the same value. That is, multiply the fraction by 1, but choose to write in a "special way."

For example, add $\dfrac{5}{36}+\dfrac{7}{54}$.

You need to find the "special numbers" that can used to create the common denominators with as small a set of numbers as possible. Why as small as possible? This will become more evident when you do an example with an algebraic fraction.

Some will immediately recognize the **_greatest common factor_** of 36 and 54, but if you don't, no problem! Just rewrite both denominators with all their prime factors: $6=2*2*3*3$, or 2^2*3^2, and $54=2*3*3*3=2*3^3$. You can see that both numbers have $2*3*3$ (or $2*3^2$) $=18$ in common.

Thus $\dfrac{5}{36}+\dfrac{7}{54}=\dfrac{5}{18*2}+\dfrac{7}{18*3}$. If you multiply the first fraction by

3/3 and the second fraction by 2/2 you get:

$$\dfrac{5}{36}+\dfrac{7}{54}=\dfrac{5}{18*2}+\dfrac{7}{18*3}=\dfrac{5}{18*2}\dfrac{3}{3}+\dfrac{7}{18*3}\dfrac{2}{2}$$

$$=\dfrac{15}{108}+\dfrac{14}{108}=\dfrac{29}{108}$$

Add: $\dfrac{23}{168}+\dfrac{37}{210}$

$168=2*84=2*2*42=2*2*2*21=2*2*2*3*7$

$210=2*105=2*5*21=2*5*3*7$

168 and 210 have $2*3*7=42$ as common factors.

$$\dfrac{23}{168}+\dfrac{37}{210}=\dfrac{23}{42*4}+\dfrac{37}{42*5}$$

Get equivalent fractions:

$$=\dfrac{23}{42*4}\dfrac{5}{5}+\dfrac{37}{42*5}\dfrac{4}{4}=\dfrac{115}{840}+\dfrac{148}{840}=\dfrac{263}{840}$$

Domain of Rational Expressions

A basic computational rule is that you do not divide by zero. You can use this principle to discuss the domain of rational expressions. You will need to eliminate any value of the variable that will cause the denominator to equal 0.

What is the domain of $\dfrac{3x+2}{x-8}$?

$x-8$ cannot equal 0, so the only value of x that you cannot use is 8. The domain is $x \neq 8$.

What is the domain of $\dfrac{x+2}{x^2-16}$?

x^2-16 cannot equal zero, so 4 and -4 cannot be members of the domain. The domain is x, where $x \neq \pm 4$.

The implication of these answers is that all other values of x are in the domain.

What is the domain of $\dfrac{x+2}{x^2+16}$?

There are no values of x for which $x^2 + 16$ will equal zero, so the domain for this expression is the set of real numbers.

Adding and Subtracting Rational Expressions

You apply the same principles to algebraic fractions as to constants.

Add: $\dfrac{x+3}{2x^2-3x-2} + \dfrac{x+4}{4x^2-1}$

Factor the denominators.

$$\frac{x+3}{2x^2-3x-2} + \frac{x+4}{4x^2-1} = \frac{x+3}{(2x+1)(x-2)} + \frac{x+4}{(2x+1)(2x-1)}$$

$$= \frac{x+3}{(2x+1)(x-2)}\frac{2x-1}{2x-1} + \frac{x+4}{(2x+1)(2x-1)}\frac{x-2}{x-2}$$

$$= \frac{(x+3)(2x-1)}{(2x+1)(x-2)(2x-1)} + \frac{(x+4)(x-2)}{(2x+1)(2x-1)(x-2)}$$

$$= \frac{(x+3)(2x-1)+(x+4)(x-2)}{(2x+1)(x-2)(2x-1)}$$

$$= \frac{(2x^2+5x-3)+(x^2+2x-8)}{(2x+1)(x-2)(2x-1)}$$

$$= \frac{3x^2+7x-11}{(2x+1)(x-2)(2x-1)}$$

Subtract $\dfrac{2x+1}{3x^2-2x-1}$ from $\dfrac{x+4}{x^2-2x+1}$.

Write the subtraction problem, and factor the denominators.

$$\frac{x+4}{x^2-2x+1} - \frac{2x+1}{3x^2-2x-1} = \frac{x+4}{(x-1)^2} - \frac{2x+1}{(3x+1)(x-1)}$$

Get a common denominator and equivalent fractions.

$$= \frac{x+4}{(x-1)^2} \frac{3x+1}{3x+1} - \frac{2x+1}{(3x+1)(x-1)} \frac{x-1}{x-1}$$

$$= \frac{(x+4)(3x+1)-(2x+1)(x-1)}{(x-1)^2(3x+1)}$$

$$= \frac{(3x^2+13x+4)-(2x^2-x-1)}{(x-1)^2(3x+1)}$$

$$= \frac{3x^2+13x+4-2x^2+x+1}{(x-1)^2(3x+1)}$$

$$= \frac{x^2+14x+5}{(x-1)^2(3x+1)}$$

Don't try to expand the denominator. You should examine the numerator to see whether it factors and whether the fraction will reduce. In this example, the numerator does not factor, so the problem is finished.

Multiplying and Dividing Rational Expressions

If you take a pizza and divide it into 8 (equal) slices, each slice represents 1/8 of the pizza. $1 \div 8 = \frac{1}{8}$. If you take one of these slices and divide it into 2 equal pieces, what part of the original pizza is each of these new slices?

Cutting each of the original slices in two will give 16 slices. Therefore, the answer to the last question is that each smaller piece will be 1/16 of the pizza. $\frac{1}{8} \div 2 = \frac{1}{16}$. You can also look at this problem as taking 1/2 of each slice (1/8 of the pizza). Hence, $\frac{1}{8} \times \frac{1}{2} = \frac{1}{16}$.

In each case, the 1/8 represents how much pizza, and the $\div 2$ or $* 1/2$ represents the action performed upon the pizza. For this reason, a common denominator is *not* needed.

Multiply: $\frac{15}{64} * \frac{8}{25}$

The basic rules for multiplication are that numerators are multiplied together and denominators are multiplied together. A consequence of this is that the magnitude of the numbers can get very large. To help you get the "more reasonable reduced" answer, factor the numerator and the denominator.

$$\frac{15}{64}*\frac{8}{25}=\frac{3*5}{8*8}*\frac{8}{5*5}=\frac{3*5*8}{8*8*5*5}=\frac{3}{8*5}*\frac{5*8}{5*8}=\frac{3}{40}$$

(Yes, you remember a quicker way to reduce this problem. But before you do that, be sure you understand that what you are about to do is really the same process as the one we just described—without the need to put all the factors into one fraction. The only operation in the numerator and in the denominator is multiplication, and the concept of "canceling" is really division of the numerator and denominator by the same number. You will find that you won't struggle with this in multiplication and division problems as you might when facing an addition or subtraction problem.)

Extending this practice to algebraic expressions, we find that the problem

$$\frac{(x-3)(x+7)}{(x+7)(x-5)}*\frac{(x-5)(5x+18)}{(x-3)(x-9)}$$

reduces to

$$\frac{(x-3)(x+7)}{(x+7)(x-5)}*\frac{(x-5)(5x+18)}{(x-3)(x-9)}=\frac{5x+18}{x-9}$$

Simplify: $\left(\frac{8x^2-14x-15}{12x^2+17x+6}\right)\left(\frac{6x^2+25x+14}{6x^2-7x-20}\right)$

Factor each of the quadratics (guess and check, split the middle, or use the zeroes) to get $\frac{(2x-5)(4x+3)}{(3x+2)(4x+3)}*\frac{(3x+2)(2x+7)}{(2x-5)(3x+4)}$. This reduces to $\frac{(2x+7)}{(3x+4)}$.

Simplify: $\left(\dfrac{x^2-9x-10}{2x^2+x-1}\right)\div\left(\dfrac{x^2-6x-40}{4x^2-1}\right)$

Change division to multiplication, and use the reciprocal of the divisor.

$$\left(\dfrac{x^2-9x-10}{2x^2+x-1}\right)\left(\dfrac{4x^2-1}{x^2-6x-40}\right)$$

Factor each of the quadratics, reduce the fraction, and record the answer.

$$\dfrac{(x-10)(x+1)}{(2x-1)(x+1)}*\dfrac{(2x-1)(2x+1)}{(x-10)(x+4)}=\dfrac{2x+1}{x+4}$$

Complex Fractions

$\dfrac{2}{3}$ is a proper fraction because the numerator is smaller than the denominator. $\dfrac{3}{2}$ is an improper fraction because the numerator is larger than the denominator. $\dfrac{2/3}{3/2}$ is called a complex fraction because the numerator and the denominator are both fractions. A **complex fraction** is one in which either the numerator or denominator (or possibly both) contain fractional expressions.

RULE

To simplify a complex fraction, multiply both the numerator and the denominator by the common denominator of the component fractions.

Simplify: $\dfrac{2/3}{3/2}$

$$\left(\dfrac{2/3}{3/2}\right)\left(\dfrac{6}{6}\right)=\dfrac{4}{9}$$

Simplify: $\dfrac{\dfrac{2}{3}+\dfrac{5}{8}}{\dfrac{3}{4}-\dfrac{5}{12}}$

The common denominator for the four component fractions is 24.

$$\left(\dfrac{\dfrac{2}{3}+\dfrac{5}{8}}{\dfrac{3}{4}-\dfrac{5}{12}}\right)\left(\dfrac{24}{24}\right)=\dfrac{\left(\dfrac{2}{3}\right)24+\left(\dfrac{5}{8}\right)24}{\left(\dfrac{3}{4}\right)24-\left(\dfrac{5}{12}\right)24}=\dfrac{16+15}{18-10}=\dfrac{31}{8}$$

Simplify: $\dfrac{x-\dfrac{4}{x-3}}{2-\dfrac{x-2}{x-3}}$

The common denominator for the four component fractions is $x-3$.

$$\left(\dfrac{x-\dfrac{4}{x-3}}{2-\dfrac{x-2}{x-3}}\right)\left(\dfrac{x-3}{x-3}\right)=\dfrac{x(x-3)-\left(\dfrac{4}{x-3}\right)(x-3)}{2(x-3)-\left(\dfrac{x-2}{x-3}\right)(x-3)}=\dfrac{x^2-3x-4}{2x-6-(x-2)}$$

$$=\dfrac{(x-4)(x+1)}{2x-6-x+2}=\dfrac{(x-4)(x+1)}{x-4}=x+1$$

Simplify: $\dfrac{x+3+\dfrac{8}{4-x}}{x-2+\dfrac{4x-23}{x-4}}$

The common denominator for the component fractions is $x-4$. Recall that $4-x=-1(x-4)$.

$$\left(\dfrac{x+3+\dfrac{8}{4-x}}{x-2+\dfrac{4x-23}{x-4}}\right)\left(\dfrac{x-4}{x-4}\right)=\dfrac{(x+3)(x-4)+\left(\dfrac{8}{4-x}\right)(x-4)}{(x-2)(x-4)+\left(\dfrac{4x-23}{x-4}\right)(x-4)}$$

$$=\dfrac{x^2-x-12+(8)(-1)}{x^2-6x+8+(4x-23)}=\dfrac{x^2-x-20}{x^2-2x-15}=\dfrac{(x-5)(x+4)}{(x-5)(x+3)}=\dfrac{x+4}{x+3}$$

Exercises for Chapter 11

Simplify each of the following by hand. Check your answers with your calculator.

1. $\dfrac{23}{128} + \dfrac{91}{192}$

2. $\dfrac{89}{240} - \dfrac{37}{144}$

3. $\dfrac{84}{125} \times \dfrac{75}{108}$

4. $\dfrac{48}{99} \times \dfrac{132}{256}$

5. $\dfrac{56}{121} \div \dfrac{49}{154}$

6. $\dfrac{99}{225} \div \dfrac{187}{400}$

7. $\dfrac{\dfrac{5}{6} - \dfrac{7}{18}}{\dfrac{7}{9} - \dfrac{2}{3}}$

8. $\dfrac{2x}{x^2 - 5x + 6} + \dfrac{x+1}{2x^2 - 5x - 3}$

9. $\dfrac{x+2}{2x^2 + x - 1} + \dfrac{x-3}{2x^2 + 5x + 3}$

10. $\dfrac{4x+3}{5x^2 + 12x + 4} + \dfrac{9x-2}{25x^2 - 4}$

11. $\dfrac{3x-2}{7x^2+3x-4}+\dfrac{9x-2}{3x^2-2x-5}$

12. $\dfrac{2x-3}{x^2-5x+6}-\dfrac{x-1}{2x^2-5x-3}$

13. $\dfrac{3x-2}{2x^2+x-1}-\dfrac{x+5}{2x^2-7x+3}$

14. $\dfrac{4x+3}{5x^2+12x+4}-\dfrac{3x-4}{4-25x^2}$

15. $\dfrac{4x-15}{9x^2-4x-5}-\dfrac{11x-7}{7x^2-2x-5}$

16. $\left(\dfrac{6x^2+7x-3}{8x^2+6x-9}\right)\left(\dfrac{12x^2-5x-3}{9x^2+12x-5}\right)$

17. $\left(\dfrac{4x^2-25}{6x^2+19x+10}\right)\left(\dfrac{9x^2-4}{6x^2-17x+5}\right)$

18. $\left(\dfrac{14x^2-31x+15}{8x^2-14x+3}\right)\div\left(\dfrac{21x^2-22x+5}{8x^2-22x+5}\right)$

19. $\left(\dfrac{4x^2-4x-15}{6x^2-x-15}\right)\div\left(\dfrac{6x^2-11x-10}{6x^2-17x-14}\right)$

20. $\dfrac{x+1+\dfrac{x-13}{x-5}}{x+2+\dfrac{2x-20}{x-5}}$

CHAPTER 12

Absolute Value and Irrational Expressions

The production of ball bearings is said to be operating within limits if the diameter of the bearings produced is within 0.02 millimeter (mm) of the design value of 1.5 mm. What is the set of values for which the process is within limits? How much time does it take for an object to fall when released from the top of a building? How much time does it take for a pendulum to complete a swing?

Absolute Value Equations and Inequalities

The **absolute value** of a number is the distance from that number to zero on a number line. Two vertical lines are used to represent this function. |5|, the absolute value of 5, is 5, and |-5| is also equal to 5, because both 5 and -5 are 5 units from zero on the number line.

The product, or quotient, of the absolute value of two numbers is equal to the absolute value of the product, or quotient, of these numbers. That is,

$$|a|\,|b| = |ab| \text{ and } \frac{|a|}{|b|} = \left|\frac{a}{b}\right|$$

Is this also true for addition and subtraction? It is certainly true that $|3+2| = |3| + |2|$. However, $|-3+2|$ is not equal to $|-3| + |2|$.

Solve: $|x| = 6$

What numbers are 6 units from zero? 6 and -6 meet this condition, so $x = \pm 6$.

Solve: $|3x| = 6$

In this case, $3x = \pm 6$, so $x = \pm 2$. Another way to look at his problem is to note that $|3x| = |3|\,|x|$ and $|3| = 3$. Thus the equation becomes $3|x| = 6$, or $|x| = 2$. Therefore, $x = \pm 2$.

Solve: $|x+2| = 6$

In this case, $x+2 = \pm 6$. Solve each equation, $x+2=6$ and $x+2=-6$, to find $x = 4, -8$. Note that -2 is midway between 4 and -8 and that -2 is the value that makes $x+2$ equal zero. The difference between this problem and the last is that the number is shifted 2 to the left. From a geometric perspective, to solve the equation $|x+2| = 6$ is to find those points that are 6 units from -2.

Solve: $|3x+2| = 6$

Here $3x+2=\pm6$. Solve each equation, $3x+2=6$ and $3x+2=\text{-}6$, to get $x=4/3$, -8/3. A second way to do this problem is to factor the 3 from the left side of the equation, $3\ |x+2/3|=6$. Divide by 3 to get $|x+2/3|=2$. The solution to this problem is those points that are 2 units from -2/3. The point 2 units to the right of -2/3 is 4/3, and the point 2 units to the left of -2/3 is -8/3.

Solve: $|8x-20|=16$

Here $8x-20=\pm16$. Solve each equation, $8x-20=16$ and $8x-20=\text{-}16$, to get $x=4.5$ and 0.5. The geometric approach to this problem is to factor out the 8, which yields $8\ |x-5/2|=16$. Divide by 8 to get $|x-5/2|=2$. The solution are those points that are 2 units from 5/2. The point 2 units to the left of 5/2 is 0.5, and the point 2 units to the right of 5/2 is 4.5.

Solve: $|x| \leq 6$

What are the points whose distance from zero is less than or equal to 6 units? The solution is those points that satisfy the inequality $\text{-}6 \leq x \leq 6$. That is, the solution is $\text{-}6 \leq x$ AND $x \geq 6$.

Solve: $|x| > 6$

The solution to this inequality will be all the points whose distance from zero is more than 6 units. That is, the solution is $x < \text{-}6$ OR $x > 6$.

Using the distance concept for solving absolute value inequalities will help you distinguish between the continuous interval $a < x < b$ and the discontinuous interval $x < a$ OR $x > b$.

Solve: $|3x-9| \geq 12$

Factor the 3 from the left-hand side, $3\ |x-3| \geq 12$, and then divide by 3, $|x-3| \geq 4$. What are the points that are *at least* 4 units from 3? Well, 4 to the left is -1, and 4 to the right is 7. Therefore, $x \leq \text{-}1$ OR $x \geq 7$.

The solution would have been exactly the same if the problem had been $|9-3x| \geq 12$. When you factor -3 from the left, within the absolute value, the inequality becomes $|-3|\,|x-3| \geq 12$ OR $3\,|x-3| \geq 12$, exactly as you just found.

The production of ball bearings is said to be operating within limits if the diameter of the bearings produced is within 0.02 millimeter (mm) of the design value of 1.5 mm. What is the set of values for which the process is within limits?

When you read this question at the beginning of the chapter, you probably said that the limits for the diameter of the bearings are between 1.48 and 1.52 mm. In terms of absolute values, the statement about limitations on the diameter can be written as $|d-1.5| \leq 0.02$.

Translate the following problem into an inequality with an absolute value. The measurement, m, for normal sugar levels in the bloodstream should be at most 20 mg/dL from 110 mg/dL.

The difference between a person's blood sugar level and 110 mg/dL should not be greater than 20. Written with absolute value, $|m-110| \leq 20$.

The Square Root Function

The square root function, $f(x)=\sqrt{x}$, is the inverse of the quadratic function $g(x)=x^2$ when the domain of $g(x)$ is restricted to $x \geq 0$. The domain of the square root function is $x \geq 0$, and the range of the function is $y \geq 0$. Being able to determine the domain of a function defined by a square root function is an important skill.

RULE

To determine the domain of a square root function, solve the inequality that states that the expression within the radical is greater than or equal to ($\geq$) 0.

Determine the domain of $k(x)=\sqrt{3x-2}$.

Set $3x-2 \geq 0$ to find $x \geq 2/3$.

Determine the domain of $p(x)=\sqrt{15-3x}$.

Set $15 - 3x \geq 0$ to find $x \leq 5$. (Did you remember to change the direction of the inequality when you solved the problem?)

Determine the domain of $q(x) = \sqrt{x^2 - 3x - 4}$.

A good way to solve $x^2 - 3x - 4 \geq 0$ is to graph the quadratic equation $y = x^2 - 3x - 4$ and then to find the points at which the graph crosses the x-axis and use the intervals for when the graph is above the x-axis. The domain of $q(x) = \sqrt{x^2 - 3x - 4}$ is $x \leq$ -1 or $x \geq 4$.

Determine the domain of $w(t) = \sqrt{\dfrac{t+3}{t-2}}$.

Use the graph of the equation $y = \dfrac{x+3}{x-2}$ on your graphing calculator to determine that the domain of $w(t)$ is $t \leq$ -3 or $t > 2$.

Irrational Functions with Higher Indices

The square root function is the inverse of the polynomial $y = x^2$. All of the polynomials whose equation is $y = x^n$, where n is a positive integer, have inverses. Take a moment to use your graphing calculator to look at the graphs of $y = x^4$, $y = x^6$, and $y = x^8$. Do you see that they are very similar to the graph of $y = x^2$? They are symmetric to the y-axis, they pass through the origin, and they never drop below the x-axis. Because they fail the horizontal-line test for 1-1 functions, you must restrict the domain to be $x \geq 0$ in order to have an inverse function.

Take a look at the graphs of $y = x^3$, $y = x^5$, and $y = x^7$. These graphs are not symmetric to the y-axis, they pass through the origin, they do have output values that are negative, and they do pass the horizontal-line test. These functions do have inverses as they exist, so the domain does not have to be restricted.

ESSENTIAL

The domain of a radical function is $x \geq 0$ when the index is even and is all real numbers when the index is odd.

The inverse of the function $f(x) = x^n$ is $g(x) = \sqrt[n]{x}$. The exponent n in the polynomial function is called the **index** in the radical function. You know that $\sqrt{16} = 4$ because $4^2 = 16$. In the same way, $\sqrt[4]{16} = 2$ because $2^4 = 16$. You also know that there are no real numbers for which $x^4 = -16$, so $\sqrt[4]{-16}$ is not defined. This is consistent with what you saw in your graph of $y = x^4$ and the restricted domain. Because -16 is not in the range of $y = x^4$, -16 cannot be in the domain of $g(x) = \sqrt[4]{x}$.

$\sqrt[3]{8} = 2$ because $2^3 = 8$, and $\sqrt[3]{-8} = -2$ because $(-2)^3 = -8$. In like manner, $\sqrt[5]{243} = 3$ because $3^5 = 243$, and $\sqrt[5]{-243} = -3$ because $(-3)^5 = -243$.

Fractional Exponents

A property of exponents is $(b^m)^n = b^{mn}$; that is, a term with an exponent raised to another exponent is the equivalent of the term raised to the product of the exponents. This rule is used to explain the meaning of fractional exponents.

RULE

The fractional exponent 1/n means an nth root: $x^{1/n} = \sqrt[n]{x}$.

For what value of n will $(x^2)^n = x$ be a true statement?

The unwritten exponent for x on the right-hand side of the equation is 1. Setting the exponents equal to each other, $2n = 1$, shows that $n = 1/2$. What does this tell you? If you let $x = 5$, the problem becomes $(5^2)^{1/2} = 5$, or $(25)^{1/2} = 5$. You know that $\sqrt{25} = 5$. You can show that for all non-negative values of x, the same result will occur, the 1/2 exponent means the same as the square root.

Does this work for any other exponent? Does the 1/3 exponent mean the same as the cube root, and does the 1/4 exponent mean the same as the fourth root? Use your calculator to examine this. Pick any value of x that you would like. For instance, try $x = 17$. Enter into your calculator $(17^3)^{(1/3)}$

—the parentheses are *very* important—and press ENTER. Try $(-17^3)^{(1/3)}$ and then $(17^4)^{(1/4)}$. In each case, you should get the number you started with as an answer.

What happens if you attempt to calculate $\sqrt{x^2}$ with a negative value of x? For example, what is the result when $\sqrt{(-5)^2}$ is entered into the calculator? (Yes, the parentheses are necessary to make sure that the calculator performs the operations in the correct order!) The result is 5, the **absolute value** of -5. That is, $\sqrt{x^2} = |x|$.

Simplify: $(16)^{3/4}$

You know that the 1/4 exponent means fourth root. What does the 3 in the numerator do to the problem? Using the property of exponents, rewrite the problem as $(16^{1/4})^3$. This simplifies to $2^3 = 8$.

Simplify: $\left(\dfrac{27}{125}\right)^{-2/3}$

Proceeding in the same manner yields

$$\left(\frac{27}{125}\right)^{-2/3} = \left(\left(\frac{27}{125}\right)^{1/3}\right)^{-2} = \left(\frac{3}{5}\right)^{-2} = \left(\frac{5}{3}\right)^2 = \frac{25}{9}.$$

Simplify: $\left(\dfrac{54x^6 y^8}{24x^{-4}y^{12}}\right)^{3/2}$

Begin by simplifying the terms inside the parentheses.

$\left(\dfrac{54x^6 y^8}{24x^{-4}y^{12}}\right)^{3/2} = \left(\dfrac{9x^{10}}{4y^4}\right)^{3/2}$. Break down the fraction, and then the terms

within the fraction, to get: $\left(\dfrac{9x^{10}}{4y^4}\right)^{3/2} = \dfrac{\left(9x^{10}\right)^{3/2}}{\left(4y^4\right)^{3/2}} = \dfrac{9^{3/2}\left(x^{10}\right)^{3/2}}{4^{3/2}\left(y^4\right)^{3/2}}$. Simplifying

each of these terms gives the answer $\dfrac{27x^{15}}{8y^6}$.

Simplifying Irrational Expressions

Recall two important properties of exponents: $(ab)^n = a^n b^n$ (when a product of terms is raised to a power, the answer is equal to the product of the terms each raised to a power) and $\left(\dfrac{a}{b}\right)^n = \dfrac{a^n}{b^n}$ (when a quotient of terms is raised to a power, the answer is equal to the quotient of the terms each raised to a power).

A consequence of these rules is that when the exponents are fractions,

$$\sqrt[n]{ab} = \sqrt[n]{a}\,\sqrt[n]{b} \text{ and } \sqrt[n]{\dfrac{a}{b}} = \dfrac{\sqrt[n]{a}}{\sqrt[n]{b}}$$

Simplify: $\sqrt{8}$

Because this is a square root function, you want to consider those factors of 8 that are squares; 4 and 2 work. By the property of exponents, $\sqrt{8}$ becomes $\sqrt{4}\,\sqrt{2} = 2\sqrt{2}$.

RULE

When simplifying $\sqrt[n]{x}$, you must remove all factors of x that are perfect nth powers.

Simplify: $\sqrt{432}$

432 is a relatively big number, and finding the largest perfect square factor might not be something you can do. You do not have to get the largest factor the first time; you just need to be sure that your final answer does not have any perfect square factors (other than 1, of course). Because $4+3+2=9$, we know that 9 is a factor of 432. $432 = 9*48$. 4 is a factor of 48, so $432 = 9*4*12$. And 4 is also a factor of 12, so $432 = 9*4*4*3$. Therefore, $\sqrt{432} = \sqrt{9*4*4*3} = 3*2*2\sqrt{3} = 12\sqrt{3}$.

Simplify: $\sqrt[3]{108}$

You are looking for perfect cube factors of 108 because the index is a 3. $108 = 27 * 4$, so $\sqrt[3]{108} = \sqrt[3]{27 * 4} = 3\sqrt[3]{4}$.

Simplify $\sqrt{\dfrac{72x^3 y^5}{50x^9 y}}$. Assume the variables represent positive numbers.

Reduce the fraction to get $\sqrt{\dfrac{36y^4}{25x^6}}$. Because each of the terms is a perfect square, the result contains no square roots. $\sqrt{\dfrac{72x^3 y^5}{50x^9 y}} = \dfrac{6y^2}{5x^3}$.

Arithmetic of Irrational Expressions

The arithmetic of irrational expressions is just like the arithmetic of algebraic expressions. Like terms can be added to and subtracted from simpler expressions, but unlike terms cannot. For example, $3x + 5x = 8x$, but $3x + 5y$ cannot be simplified. Like and unlike terms can be multiplied and divided. For example, $(3x)(5x) = 15x^2$, and $(3x)(5y) = 15xy$. $\dfrac{15x^2}{5x^2} = 3$, $\dfrac{15x^2}{5x} = 3x$, and $\dfrac{15x^2 y^3}{5xy^5} = \dfrac{3x}{y^2}$.

Similarly, $2\sqrt{3} + 6\sqrt{3} = 8\sqrt{3}$, but $2\sqrt{3} + 3\sqrt{6}$ cannot be simplified because the **radicands**—the terms under the radical—are different. $\left(2\sqrt{3}\right)\left(6\sqrt{3}\right) = 12\sqrt{9} = 12 \cdot 3 = 36$ and $\left(2\sqrt{3}\right)\left(3\sqrt{6}\right) = 6\sqrt{18} = 6\sqrt{9}\sqrt{2} = 18\sqrt{2}$.

Simplify: $8\sqrt{5} + 3\sqrt{20} - \sqrt{500}$

At first look you might think that these terms cannot be combined because the radicands are different. However, if you simplify $3\sqrt{20}$ to be $3\sqrt{4}\sqrt{5} = 3 \cdot 2\sqrt{5} = 6\sqrt{5}$ and $\sqrt{500}$ to be $\sqrt{100}\sqrt{5} = 10\sqrt{5}$, the problem now becomes to simplify to $8\sqrt{5} + 6\sqrt{5} - 10\sqrt{5}$, which equals $4\sqrt{5}$.

Simplify: $15\sqrt{8} + 20\sqrt{\dfrac{1}{2}} - 8\sqrt{50}$

$$15\sqrt{8}=15\sqrt{4}\sqrt{2}=15{\cdot}2\sqrt{2}=30\sqrt{2} \text{ and } 8\sqrt{50}=8\sqrt{25}\sqrt{2}=8{\cdot}5\sqrt{2}=40\sqrt{2}.$$

The third term in the problem is different in that there is a fraction within the square root. Simplifying the fraction makes $20\sqrt{\dfrac{1}{2}}$ become $20\dfrac{\sqrt{1}}{\sqrt{2}}=\dfrac{20}{\sqrt{2}}$.

Although this is correct, it does not give a form that can be simplified with the other terms. If you multiply both the numerator and the denominator by $\sqrt{2}$, you get an equivalent fraction with a rational denominator (which is why this process is called ***rationalizing the denominator***), and $\dfrac{20}{\sqrt{2}}$

becomes $\dfrac{20}{\sqrt{2}}\dfrac{\sqrt{2}}{\sqrt{2}}=\dfrac{20\sqrt{2}}{\sqrt{4}}=\dfrac{20\sqrt{2}}{2}=10\sqrt{2}$. Simplifying the terms yields

$$15\sqrt{8}+20\sqrt{\dfrac{1}{2}}-8\sqrt{50}=30\sqrt{2}+10\sqrt{2}-40\sqrt{2}=0.$$

Simplify: $6\sqrt[3]{16}+5\sqrt[3]{54}+10\left(\sqrt[3]{\dfrac{1}{4}}\right)$

These are cube roots, so you need to be thinking about perfect cubes when simplifying terms. $6\sqrt[3]{16}=6\sqrt[3]{8}\sqrt[3]{2}=6{\cdot}2\sqrt[3]{2}=12\sqrt[3]{2}$ and $5\sqrt[3]{54}=5\sqrt[3]{27}\sqrt[3]{2}=5{\cdot}3\sqrt[3]{2}=15\sqrt[3]{2}$. Rewriting 1/4 as the equivalent fraction 2/8 makes $10\left(\sqrt[3]{\dfrac{1}{4}}\right)=10\left(\sqrt[3]{\dfrac{2}{8}}\right)=10\left(\dfrac{\sqrt[3]{2}}{\sqrt[3]{8}}\right)=10\left(\dfrac{\sqrt[3]{2}}{2}\right)=5\sqrt[3]{2}$. Therefore,

$$6\sqrt[3]{16}+5\sqrt[3]{54}+10\left(\sqrt[3]{\dfrac{1}{4}}\right)=12\sqrt[3]{2}+15\sqrt[3]{2}+5\sqrt[3]{2}=32\sqrt[3]{2}.$$

To rationalize the denominator of a term in which the denominator is a binomial, such as $\dfrac{8}{1+\sqrt{3}}$, multiply the numerator and denominator by the conjugate of the given denominator, thus using the formula for the difference of two squares.

Rationalize the denominator of $\dfrac{8}{1+\sqrt{3}}$, and simplify the fraction.

$$\frac{8}{1+\sqrt{3}}\left(\frac{1-\sqrt{3}}{1-\sqrt{3}}\right)=\frac{8-8\sqrt{3}}{1^2-\left(\sqrt{3}\right)^2}=\frac{8-8\sqrt{3}}{1-3}=\frac{8-8\sqrt{3}}{-2}=4\sqrt{3}+4$$

Solving Equations with Irrational Expressions

The time it takes an object to fall to the ground in a free-fall situation is given by the equation $t(h)=\sqrt{\dfrac{h}{16}}$, where h is in feet and t is in seconds.

Determine the time it takes a ball to reach the ground if it is dropped from the top of a building 120 feet tall.

$$t(120)=\sqrt{\frac{120}{16}}=2.739 \text{ seconds}$$

The period of a pendulum, the amount of time needed for it to swing back and forth, is given by the formula $T(L)=2\pi\sqrt{\dfrac{L}{9.8}}$, where L is the length of the pendulum in meters.

Determine the period of a pendulum whose length is 0.5 meter.

$$T(0.5)=2\pi\sqrt{\frac{0.5}{9.8}}=1.419 \text{ seconds}$$

The period of a pendulum is 1.1 seconds. Find, to the nearest hundredth of a meter, the length of the pendulum.

The period T is 1.1 seconds. Substituting this into the equation for the period gives $1.1=2\pi\sqrt{\dfrac{L}{9.8}}$. To solve for L, first divide by 2π to get $\dfrac{1.1}{2\pi}=\sqrt{\dfrac{L}{9.8}}$. Square both sides of the equation to remove the square root function: $\left(\dfrac{1.1}{2\pi}\right)^2=\dfrac{L}{9.8}$. Multiply by 9.8 to solve for L to determine that the length of the pendulum is 0.27 meter.

When solving equations with radicals, the process requires that you isolate a radical so that it can be removed by raising both sides of the equation to an appropriate power.

Solve: $\sqrt{2x+13} - 5 = 4$

Add 5 to get $\sqrt{2x+13} = 9$. Square both sides of the equation to get $2x+13=81$. Subtract 13 and divide by 2 to determine that $x=34$.

Solve: $\sqrt[3]{2x+13} - 5 = 4$

Add 5 to both sides of the equation: $\sqrt[3]{2x+13} = 9$. Cube both sides of the equation: $2x+13=729$. Subtract 13 and divide by 2 to determine that $x=358$.

Solve: $\sqrt{2x+13} - 5 = x$

Add 5 to both sides to get $\sqrt{2x+13} = x+5$. Square both sides to get $2x+13 = (x+5)^2$. Expand the right-hand side to get $2x+13 = x^2 + 10x + 25$. Subtract $2x+13$ from both sides of the equation: $0 = x^2 + 8x + 12$. Factor and solve: $(x+2)(x+6) = 0$, so $x=$-2, -6. Check these values into the original equation. When $x=$-2, $\sqrt{2x+13} - 5$ becomes $\sqrt{2(-2)+13} - 5$ $= \sqrt{-4+13}$ -5 $= \sqrt{9} - 5 = 3 - 5 = $-2. When $x=$-6, $\sqrt{2x+13} - 5$ becomes $\sqrt{2(-6)+13} - 5 = \sqrt{-12+13} - 5 = \sqrt{1} - 5 = 1 - 5 = $-4, not -6. Therefore, $x=$-6 must be rejected as a solution. The solution to this problem is $x=$-2.

Solve: $\sqrt{3x+4} + \sqrt{x+2} = 8$

Be warned: Because the process for solving equations with radicals is to isolate a radical and then raise both sides of an equation to an appropriate power, you will have to isolate one of the radical equations but will not lose the other.

Subtract $\sqrt{x+2}$ from both sides of the equations to get $\sqrt{3x+4}=8-\sqrt{x+2}$. Square both sides of the equation: $\left(\sqrt{3x+4}\right)^2=\left(8-\sqrt{x+2}\right)^2$. The equation becomes:

$$3x+4=64-16\sqrt{x+2}+x+2$$

Gather like terms on the right-hand side.

$$3x+4=x+66-16\sqrt{x+2}$$

Isolate the radical.

$$2x-62=\text{-}16\sqrt{x+2}$$

Square both sides of the equation.

$$(2x-62)^2=(\text{-}16\sqrt{x+2})^2$$
$$4x^2-248x+3844=256(x+2)$$
$$4x^2-248x+3844=256x+512$$
$$4x^2-504x+3332=0$$

Divide by 4.

$$x^2-126x+833=0$$

Factor and solve, or use the quadratic formula, to find $x=7, 119$. Check these answers in the original equations to find that 119 must be rejected and 7 is accepted. The solution is $x=7$.

Exercises for Chapter 12

Simplify each of the following.

1. $|\text{-17}|$

2. $|(-3)(-5)|$

3. $\left|\dfrac{x-3}{3-x}\right|$

4. $|x^2+9|$

Simplify each of the following. (Assume that all variables represent positive values.)

5. $\sqrt{75}$

6. $\sqrt{52}$

7. $\sqrt[3]{72}$

8. $\sqrt[5]{96}$

9. $\sqrt{16x^4y^6}$

10. $\sqrt{75x^3y^5}$

11. $\sqrt[3]{27x^6y^{12}}$

12. $\sqrt[3]{-64x^4y^{10}}$

13. $8\sqrt{12}+6\sqrt{27}-4\sqrt{48}$

14. $8\sqrt{18}+10\sqrt{50}-40\sqrt{\dfrac{1}{8}}$

15. $4\sqrt[3]{16}+5\sqrt[3]{54}-\dfrac{2}{5}\sqrt[3]{2000}$

16. $(25x^4)^{3/2}$

17. $\left(\dfrac{189x^6}{84y^4}\right)^{-3/2}$

Solve each of the following.

18. $|x+5|=9$

19. $|x-11|=4$

20. $|4x+5|=8$

21. $|5x-4|=11$

22. $|x-2|>3$

23. $|p+6| \leq 4$

24. $|3v+5| < 7$

25. $|2c-9| \geq 5$

26. Translate into an absolute value inequality: "All ticket prices for the play are within $25 of the average price of $60."

Find the domain for each of the following functions.

27. $f(x)=\sqrt{5x+4}$

28. $g(x)=\sqrt[3]{3x-8}$

29. $k(x)=\sqrt{x^2-4x-5}$

30. $p(x)=\sqrt{\dfrac{2x-5}{x-6}}$

Rationalize the denominator for each fraction.

31. $\dfrac{18}{\sqrt{12}}$

32. $\dfrac{72}{\sqrt[3]{16}}$

33. $\dfrac{12}{3+\sqrt{5}}$

Solve each of the equations.

34. $\sqrt{3x-2}=7$

35. $\sqrt[3]{5x-1}=4$

36. $\sqrt[5]{7x+5}=3$

37. $\sqrt{5x+1}+1=x$

38. $\sqrt{6x+31}+4=x$

39. $\sqrt{8x-23}=2x-11$

CHAPTER 13

Complex Numbers

A number of applications in physics and engineering use complex numbers. Among these are digital signal processing, fluids, stress, and image processing. Fractals, a branch of geometry made popular by Benoît Mandelbrot in the 1960s, uses functions of complex numbers to describe chaotic phenomena such as weather systems.

Extending the Number System

Think about how you learned numbers. As a youngster, you held up as many fingers as your age when someone asked how old you were. You learned to count to ten and then learned about numbers with funny names: eleven, twelve. (Wouldn't it have made more sense to call them oneteen, twoteen, and threeteen, because the next number is fourteen? And why is it fifteen and not fiveteen?) Depending on where you live, you may have learned about negative numbers in the winter when the temperature dropped below zero. Fractions were not hard to understand; you know what half a cake looks like. You may have found the rules for the arithmetic of fractions difficult to understand, but the idea of fractions themselves probably made sense.

Square roots probably make sense from an arithmetic perspective rather than a physical one. It's hard to think of $\sqrt{2}$ pieces of cake. Still, you understand that a negative times a negative is positive, just as is the product of two positive numbers, and you understand that in order to have a square root "function," the negatives had to be eliminated as a value of the *radicand* (see Chapter 12).

We hope that you follow and agree with what is written in the previous two paragraphs. But here is the big question: Can you show your friend a 1? Not a pen, not a finger—a 1. No, you can't. Numbers, like words, are mental abstractions that are used to explain and describe the world. The numbers that we call the real numbers can have a physical representation. That is, it is possible to construct a segment whose length is $\sqrt{2}$ times some length that has been designated as a unit length, a length of 1. It is also true that every one of the real numbers has a unique place on a number line once the origin is marked and a unit length determined. For these reasons, we can say that $\sqrt{-1}$ is not a real number, but it is a number. Unfortunately, the name given to this number and its multiples is ***imaginary numbers***. By definition, $\sqrt{-1} = i$.

One of the first things you have to deal with is that the imaginary unit looks like a "letter," not a number. After all, it is not made from a combination of the ten digits with which all the real numbers are made. It will help to learn that not only is $\sqrt{-1} = i$ but, as a consequence, $i^2 = -1$. Multiply both sides of this equation by i and you have $i^3 = -i$. Multiply by i again and you have $i^4 = 1$. Do you see that i^5 through i^8 will give the answers i, -1, -i, 1 again?

Powers of *i* repeat in a cycle of four elements.

Evaluate i^{275}.

Because i^4, i^8, i^{12}, and so on all equal 1, you need to divide 275 by 4. The remainder, if there is a remainder, will tell you the answer. If there is no remainder, then the answer is 1. The remainder when 275 is divided by 4 is 3. Therefore, $i^{275} = i^3 = \text{-}i$.

When simplifying square roots of negative numbers, always factor out the square root of -1 as the first step.

Simplifying radicals with negative radicands requires a little more care.

Simplify: $\sqrt{\text{-}25}$

$$\sqrt{\text{-}25} = \sqrt{\text{-}1}\sqrt{25} = (i)(5) = 5i$$

Simplify: $\sqrt{\text{-}75}$

$$\sqrt{\text{-}75} = \sqrt{\text{-}1}\sqrt{25}\,\sqrt{3} = 5i\,\sqrt{3}$$

Simplify: $\left(\sqrt{\text{-}16}\right)\left(\sqrt{\text{-}9}\right)$

The temptation is to multiply -16 and -9, get 144, take the square root, and call the answer 12. *This is wrong.* You must first simplify the radicals, $\sqrt{\text{-}16} = 4i$ and $\sqrt{\text{-}9} = 3i$ and the product of $4i$ and $3i$ is $12i^2$. Because $i^2 = \text{-}1$, $12i^2 = \text{-}12$. This is the correct answer.

Complex Numbers: A Mix of the Real and the Imaginary

If i is a number but is not on the number line, how can it be represented? Instead of thinking of a number line, think of a number plane with the real numbers on the horizontal axis and the imaginary numbers on the vertical axis. The axes intersect at 0, the origin for both systems. (There's one bit of good news: 0 times i is 0, so multiplication by 0 has not been changed.)

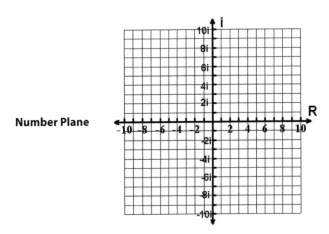

Number Plane

The real numbers are on the horizontal axis and the imaginary numbers are on the vertical axis. That's simple. What about the numbers in the "quadrants"? For example, what do you call the number that is 3 to the right of the origin and four above the real axis? This is a bit more complex. The number is written as $3 + 4i$. All other numbers with real component a and imaginary component b are written as $a + bi$ and are called ***complex numbers***. If $a = 0$, the number is imaginary, and if $b = 0$, the number is real. The only number common to the sets of the real and the imaginary numbers is 0. Each of the sets of the real and the imaginary numbers is a subset of the complex numbers.

Complex numbers are represented as directed segments that start at the origin. Don't let the arrow at the end of the directed segment lead you to believe that the figure is a ray; the arrow simply indicates the direction of flow. (You will later learn to think of this as a vector.) The graph of $3 + 4i$ looks like this:

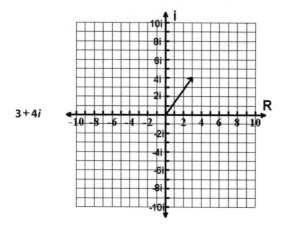

$3+4i$

The graph of $-4+5i$ looks like this:

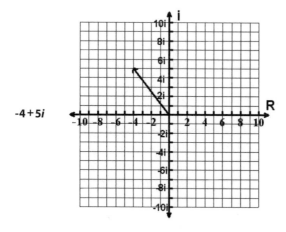

$-4+5i$

Adding and Subtracting Complex Numbers

The arithmetic of complex numbers uses the same rules as the arithmetic of binomials. When adding or subtracting, combine like terms.

ESSENTIAL

Complex numbers are always written in the form $a + bi$, with the real term first and the imaginary term second.

Simplify: $(3+4i)+(-4+5i)$

$3+(-4)=-1$ and $4+5=9$, so $(3+4i)+(-4+5i)=-1+9i$.

Simplify: $(3+4i)-(-4+5i)$

$3-(-4)=7$ and $4-5=-1$, so $(3+4i)-(-4+5i)=7-i$.

Multiplying Complex Numbers

Use the distributive property when multiplying complex numbers, and remember to simplify powers of i.

Simplify: $(3+4i)(-4+5i)$

$3(-4+5i)=-12+15i$ and $4i(-4+5i)=-16i+20i^2$, so $(3+4i)(-4+5i)=$ $-12+15i-16i+20i^2$. Because $20i^2=-20$, you can combine like terms to get $-32-i$ as the product.

Simplify: $(5+2i)(6-7i)$

$5(6-7i)=30-35i$ and $2i(6-7i)=12i-14i^2$, so $(5+2i)(6-7i)=30-35i+12i-14i^2$. Because $-14i^2=14$, combine like terms to get $44-23i$ as the product.

Simplify: $(5+12i)^2$

$(5+12i)^2=5^2+2(5)(12i)+(12i)^2=25+120i+144i^2=25+120i-144$ $=-119+120i$

Simplify: $\left(5+\sqrt{-12}\right)\left(8-\sqrt{-27}\right)$

Simplify the radicals first.

$$5+\sqrt{-12}=5+\sqrt{-1}\sqrt{4}\sqrt{3}=5+2i\sqrt{3}$$

and

$$8-\sqrt{-27}=8-\sqrt{-1}\sqrt{9}\sqrt{3}=8-3i\sqrt{3}$$

$$\left(5+\sqrt{-12}\right)\left(8-\sqrt{-27}\right)=\left(5+2i\sqrt{3}\right)\left(8-3i\sqrt{3}\right)$$
$$=40-15i\sqrt{3}+16i\sqrt{3}-6i^2\left(\sqrt{3}\right)^2=40+i\sqrt{3}-6(-1)(3)$$
$$=40+i\sqrt{3}+18=58+i\sqrt{3}$$

Complex Conjugates

Just as $a+b$ and $a-b$ are conjugates with $(a+b)(a-b)=a^2-b^2$, $a+bi$ and $a-bi$ are called **complex conjugates**.

$$(a+bi)(a-bi)=a^2+b^2$$

You now have to be more careful when you speak and write. In the past, you would have said that the sum of two squares did not factor. Now you will have to say that the sum of two squares does not factor over the real numbers, but it does factor over the complex numbers.

RULE

The product of complex conjugates will always be a real number.

Simplify: $(4+3i)(4-3i)$

$$(4+3i)(4-3i)=16-12i+12i-9i^2=16+9=25$$

Simplify: $\left(2+\sqrt{-46}\right)\left(2-\sqrt{-46}\right)$

$$\left(2+\sqrt{-46}\right)\left(2-\sqrt{-46}\right)=\left(2+i\sqrt{46}\right)\left(2-i\sqrt{46}\right)=2^2+\left(\sqrt{46}\right)^2=4+46=50$$

Dividing Complex Numbers

Division of complex numbers is accomplished by rewriting the problem as a fraction and then getting an equivalent fraction by multiplying the numerator and denominator of the fraction by the conjugate of the denominator.

Simplify: $\dfrac{4+3i}{2-i}$

Multiply numerator and denominator by $2+i$.

$$\frac{4+3i}{2-i}\cdot\frac{2+i}{2+i}=\frac{8+10i+3i^2}{4+1}=\frac{5+10i}{5}=1+2i$$

Simplify: $\dfrac{4-\sqrt{-18}}{3+\sqrt{-8}}$

Simplify the radicals.

$$\frac{4-\sqrt{-18}}{3+\sqrt{-8}}=\frac{4-\sqrt{-1}\sqrt{9}\sqrt{2}}{3+\sqrt{-1}\sqrt{4}\sqrt{2}}=\frac{4-3i\sqrt{2}}{3+2i\sqrt{2}}$$

Multiply by the conjugate of the denominator.

$$\frac{4-3i\sqrt{2}}{3+2i\sqrt{2}}\cdot\frac{3-2i\sqrt{2}}{3-2i\sqrt{2}}=\frac{12-17i\sqrt{2}+6i^2\left(\sqrt{2}\right)^2}{3^2+\left(2\sqrt{2}\right)^2}=\frac{0-17i\sqrt{2}}{17}=-i\sqrt{2}$$

Simplify: $\dfrac{10+7i\sqrt{6}}{3+2i\sqrt{6}}$

Multiply by the conjugate of the denominator.

$$\frac{10+7i\sqrt{6}}{3+2i\sqrt{6}}\cdot\frac{3-2i\sqrt{6}}{3-2i\sqrt{6}}$$

Expand the numerator and denominator.

$$\frac{10+7i\sqrt{6}}{3+2i\sqrt{6}}\cdot\frac{3-2i\sqrt{6}}{3-2i\sqrt{6}}=\frac{30+i\sqrt{6}-14i^2\left(\sqrt{6}\right)^2}{3^2+\left(2\sqrt{6}\right)^2}$$

Simplify.

$$\frac{10+7i\sqrt{6}}{3+2i\sqrt{6}}\cdot\frac{3-2i\sqrt{6}}{3-2i\sqrt{6}}=\frac{30+i\sqrt{6}-14i^2\left(\sqrt{6}\right)^2}{3^2+\left(2\sqrt{6}\right)^2}=\frac{30+i\sqrt{6}+84}{9+24}=\frac{114+i\sqrt{6}}{33}=\frac{114}{33}+\frac{\sqrt{6}}{33}i$$

Simplify: i^{-7}.

$i^{-7}=\dfrac{1}{i^7}$. Multiply numerator and denominator by i gives $\dfrac{i}{i^8}$. Because $i^8=1$, $i^{-7}=i$.

The Discriminant and the Nature of the Roots of a Quadratic Equation

The solution to the quadratic equation $ax^2+bx+c=0$ is

$$x=\frac{-b\pm\sqrt{b^2-4ac}}{2a}$$

The radicand for this equation, b^2-4ac, is called the **_discriminant_**, because it discriminates among the various sets of numbers from which the solutions may be drawn.

b^2-4ac	Nature of the Roots	Example
<0	The roots are complex conjugates.	$8x^2-12x+5=0$
$=0$	The roots are real, rational, and equal.	$4x^2-20x+25=0$
>0 and a perfect square	The roots are real, rational, and unequal.	$16x^2-16x+3=0$
>0 and _not_ a perfect square	The roots are real, irrational, and unequal.	$6x^2-5x-3=0$

Find the discriminant, and determine the nature of the roots for the equation $5x^2+3x+2=0$.

The discriminant is $3^2 - 4(5)(2) = -31$. Therefore, the roots are complex conjugates.

Find the discriminant, and determine the nature of the roots for the equation $5x^2 + 3x - 2 = 0$.

The discriminant is $3^2 - 4(5)(-2) = 49$. Therefore, the roots are real, rational, and unequal.

Find the discriminant, and determine the nature of the roots for the equation $6x^2 + 3x - 2 = 0$.

The discriminant is $3^2 - 4(6)(-2) = 57$. Therefore, the roots are real, irrational, and unequal.

Solving Quadratic Equations

There was a time when you would respond "no real solutions" when the radicand from the quadratic formula was negative, or when one side of a quadratic equation was negative and the other side was not, or when the graph of the corresponding parabola did not cross the x-axis. Now you need to realize that the solutions to equations with these results come from the set of complex numbers.

Solve: $81x^2 - 108x + 37 = 0$

Look at the discriminant, $(-108)^2 - 4(81)(37) = -324$. The roots are complex. Use the quadratic formula:

$$x = \frac{-(-108) \pm \sqrt{-324}}{2(81)} = \frac{108 \pm 18i}{162} = \frac{108}{162} \pm \frac{18}{162}i = \frac{2}{3} \pm \frac{1}{9}i$$

Solve by completing the square: $-3x^2 + 6x - 4 = 0$

Add 4 to both sides of the equation: $-3x^2 + 6x = 4$
Factor -3 from the left side of the equation: $-3(x^2 - 2x) = 4$

Divide by -3: $x^2 - 2x = -4/3$

Complete the square by adding 1 to both sides of the equation:

$x^2 - 2x + 1 = -1/3$

Factor the left-hand side of the equation: $(x-1)^2 = -1/3$

Note that the left-hand side of the equation is "not negative" but the right-hand side is.

Take the square root of both sides of the equation (remember the $\pm$):

$x - 1 = \pm\sqrt{-1/3}$

Solve for x: $x = 1 \pm \dfrac{1}{\sqrt{3}}i$, or $x = 1 \pm \dfrac{\sqrt{3}}{3}i$

Sum and Product of the Roots

The roots of the quadratic equation are $x = \dfrac{-b + \sqrt{b^2 - 4ac}}{2a}$ and $x = \dfrac{-b - \sqrt{b^2 - 4ac}}{2a}$. When these numbers are added together, the result is $-b/a$, and when they are multiplied, the result is c/a. This information is very useful for finding the second root of a quadratic equation when given the first root and for writing a quadratic equation when given its roots.

One root of the quadratic equation $6x^2 - 11x - 72 = 0$ is 9/2. Find the other root.

Calling the second root of the equation r, you can use the product of the roots to write the equation $(9/2)r = -72/6$. Divide by 9/2 to find that $r = -8/3$. You could also have used the sum of the roots to solve the equation $9/2 + r = 11/6$.

Write a quadratic equation with integral coefficients whose roots are 5/8 and -9/4.

"Integral coefficients" means that the coefficients must be integers, not fractions, and "Write a quadratic equation" means that $=0$ needs to be part of your answer.

The sum of the roots is 5/8 + -9/4 = -13/8 = -b/a. The product of the roots is (5/8)(-9/4) = -45/32 = c/a. In order for this process to work, the denominators (that is, the value of a) must agree. Rewriting the sum as an equivalent fraction with denominator 32 gives -b/a = -52/32. Therefore, $a = 32$, -b = -52 so $b = 52$, and $c = -45$. The quadratic equation is $32x^2 + 52x - 45 = 0$.

Write a quadratic equation with integral coefficients whose roots are $\frac{3}{4} \pm \frac{\sqrt{5}}{2}$.

The sum of the roots is 3/2, and the product is -11/16. Rewriting the sum to be 24/16, you have that $a = 16$, $b = -24$, and $c = -11$. The quadratic equation is $16x^2 - 24x - 11 = 0$.

Write a quadratic equation with integral coefficients whose roots are $\frac{-4}{5} \pm \frac{3\sqrt{2}}{10} i$.

The sum of the roots is -8/5, and the product of the roots is 41/50. When you rewrite the sum as -80/50, you have $a = 50$, $b = 80$, and $c = 41$. The equation is $50x^2 + 80x + 41 = 0$.

Exercises for Chapter 13

Simplify each of the following.

1. i^{27}

2. i^{-5}

3. $\sqrt{-100}$

4. $\sqrt{-75}$

5. $\left(\sqrt{-12}\right)\left(\sqrt{-108}\right)$

6. $(8+7i)+(\text{-}3+4i)$

7. $(4-5i)-(9-7i)$

8. $\left(\text{-}2+5i\sqrt{3}\right)+\left(\text{-}5-2i\sqrt{3}\right)$

9. $4\sqrt{\text{-}50}+3\sqrt{\text{-}18}-2\sqrt{\text{-}72}$

10. $3\sqrt{\text{-}20}-5\sqrt{\text{-}45}+2\sqrt{\text{-}500}$

11. $(6+7i)(9-4i)$

12. $(12+5i)(12-5i)$

13. $(7+3i)^2$

14. $\dfrac{5+2i}{3-i}$

15. $\dfrac{4+\sqrt{\text{-}24}}{1-\sqrt{\text{-}6}}$

16. $\dfrac{10-9i\sqrt{5}}{1+4i\sqrt{5}}$

Determine the nature of the roots to each of the quadratic equations without solving the equations.

17. $8x^2-5x-3=0$

18. $8x^2-5x+3=0$

19. $6x^2+8x+1=0$

20. $18x^2-48x+32=0$

Solve each of the quadratic equations.

21. $8x^2 - 5x - 3 = 0$

22. $8x^2 - 5x + 3 = 0$

23. $6x^2 + 8x + 1 = 0$

24. $18x^2 - 48x + 32 = 0$

25. $-9x^2 + 10x + 5 = 0$

Write quadratic equations with integral coefficients whose roots are

26. -3/7, 2/3

27. $-3/7 \pm 2/3\, i$

28. $5 \pm 2i\sqrt{3}$

29. $\dfrac{-8}{3} \pm \dfrac{\sqrt{11}}{3}\, i$

CHAPTER 14

Transformations of Functions

New functions can be made from basic functions by sliding them to a new position on the plane, by reflecting them over the *x*-axis or the *y*-axis, or possibly by stretching or shrinking them from one of the axes. The combination of the base function and all possible transformations of this function give a family of functions.

Vertical Slides

Enter and graph the function $y=x^2$ on your calculator. Enter a new function, $y=x^2-1$. What is the relationship between the original graph and this transformed graph? Edit the transformed graph to read $y=x^2-4$ Again, what happened to the graph of the original function?

Edit the transformed function to read $y=x^2+2$. How is this transformed graph similar to the first two transformations? How is it different? Sketch $y=x^2+4$ on a piece of paper before you edit the function in your calculator. Once you have sketched the graph on paper, use your calculator to sketch this same graph. Does the graph on the calculator agree with your paper-and-pencil sketch?

Edit the original function to be $y=|x|$. Is this graph transformed in the same way as $y=x^2$ is when you graph $y=|x|-1$, $y=|x|-4$, $y=|x|+2$, and $y=|x|+4$?

At this point, you probably have determined the impact of adding a number to, or subtracting a number from, a function. When c is a positive number:

$y=f(x)-c$ will slide the graph of $f(x)$ down c units.

$y=f(x)+c$ will slide the graph of $f(x)$ up c units.

Horizontal Slides

Continue with the basic function $y=|x|$. Rewrite the second function in this exercise to be $y=|x-1|$. How has the original graph been transformed?

Rewrite the edited function to be $y=|x-4|$. Before graphing this function, write down how you think the original graph will be transformed. Having made your prediction, graph the new function. Were you correct in your prediction?

Predict how the graph of $y=|x|$ will be transformed when the equation is changed to $y=|x+2|$ and then to $y=|x+4|$.

The motions seem to be different from the motions for the vertical transformations. When c is a positive number:

$y=f(x-c)$ slides the graph of $f(x)$ to the right c units.

$y = f(x+c)$ slides the graph of $f(x)$ to the left c units.

You know that $|-1| = 1$ and that $|1| = 1$. When $y = |x - 4|$, the function will give the output value of 1 when $x - 4$ is equal to -1 or when $x - 4$ is equal to 1. Thus $x - 4 = -1$ when $x = 3$, and $x - 4 = 1$ when $x = 5$. These answers are 4 units to the right of the original values.

Describe how the graph of $y = (x - 3)^2$ is transformed from the graph of $y = x^2$.

The graph of $y = x^2$ is slid to the right 3 units.

Write an equation for the graph shown.

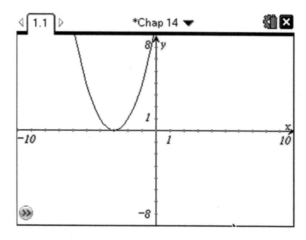

The graph is a parabola moved to the left 3 units. Its equation is $y = (x + 3)^2$.

Stretches from the Horizontal Axis

You have studied what happens when you add a number to, or subtract a number from, the output of a function ($f(x) \pm c$) or when you add a number to, or subtract a number from, the input of a function before the output is computed ($f(x \pm c)$). What happens to the graph of a function when the output is multiplied? That is, what impact does c in the equation $y=cf(x)$ have on the graph of $f(x)$ when c is a positive number?

Make your base function be $y=x^2$, and make the first edited function be $y=2x^2$. How does the graph of the new function compare to the graph of the base function? Examine the graphs of $y=4x^2$, $y=5x^2$, and $y=8x^2$. What do you observe about these graphs when you compare them to the base function $y=x^2$ and to each other?

Change the edited function to $y=1/2x^2$. Does the graph of this function change the graph of $y=x^2$ in a way that makes sense to you?

In both cases, the graph of $y=cx^2$ appears to be stretched away from the x-axis when $c > 1$ and to be stretched toward the x-axis when $0 < c < 1$. That is, the graph of $y=x^2$ is **dilated** from the x-axis.

Use your calculator to change the base function from $y=x^2$ to other functions (such as $y=|x|$, $y=x^3$, or $y=1/x$), introduce different coefficients to these functions, and be sure that you understand the impact the coefficient has on the graph of the base function.

The graph of $y=f(cx)$ is a different transformation from $y=cf(x)$. However, since you are examining polynomial, rational, and irrational functions at this time, the significance is difficult to see because of your ability to perform algebra on the function. For example, the graph of $y=(2x)^2$ can be examined by rewriting the function to be $y=4x^2$, and you can see that the output values of $y=x^2$ are dilated from the x-axis by a factor of 4.

Reflections

The impact of a negative coefficient is to reflect the graph over an axis. Let the basic function be $y=x^2$. Make the edited function $y=-x^2$. Graph both functions. You can see that the edited function is reflected over the x-axis. $y=x^2$ and $y=|x|$ are both symmetric to the y-axis and do not illustrate an important transformation. Change the base function to $y=\sqrt{x}$, and make the transformed function $y=-\sqrt{x}$. Graphing these functions illustrates once again that the graph of $y=-f(x)$ will be the reflection of $y=f(x)$ over the x-axis. The domain for each of these functions is $x \geq 0$, but the range changes. The range for $y=\sqrt{x}$ is $y \geq 0$, whereas the range for $y=-\sqrt{x}$ is $y \leq 0$.

Change the transformed function to be $y=\sqrt{-x}$. The graph of $y=\sqrt{x}$ is reflected over the y-axis, not over the x-axis. The domain of $y=\sqrt{x}$ is $x \geq 0$, whereas the range is $y \geq 0$. The domain of $y=\sqrt{-x}$ is $x \leq 0$, whereas the range stays as $y \geq 0$.

Putting It All Together

You can now put all these transformations together to better understand the families of functions. Knowing the graph of the parent functions and being able to describe the transformations will enable you to have a clear mental image of the graphs of the functions before you use your calculator.

ESSENTIAL

When describing the transformations, begin with the domain, and then work your way out of the problem (that is, work from the inside out).

Given $f(x)=4\sqrt{x-3}-2$, describe the transformations to the parent function $y=\sqrt{x}$.

Because x is replaced with $x-3$, the graph is translated to the right 3 units. The coefficient of 4 causes the graph to be stretched from the x-axis by a factor of 4. Finally, the graph is translated down by 2 because of the -2 at the end of the problem.

What are the domain and range of $f(x)=4\sqrt{x-3}-2$?

The domain and range of the parent function $y=\sqrt{x}$ are $x \geq 0$ and $y \geq 0$, respectively. Because the graph was translated 3 units to the right and 2 units down, the domain of $f(x)$ is $x \geq 3$, and the range of $f(x)$ is $y \geq -2$.

Given $g(x)=-2(x+1)^2+7$, describe the transformations to the parent function $y=x^2$.

Because x is replaced with $x+1$, the graph is translated left 2 units. The coefficient -2 causes the graph to reflect over the x-axis and to be stretched from the x-axis by a factor of 2. Adding 7 at the end of the function translates the graph up 7 units.

What are the domain and range of $g(x)=-2(x+1)^2+7$?

The domain and range of the parent function $y=x^2$ are the set of real numbers and $y \geq 0$, respectively. Translating the graph 1 unit to the left does not change the domain from the set of real numbers. Reflecting the graph over the x-axis and adding 7 changes the range to $y \leq 7$.

Given $f(x)=x^2$, write an equation for the function $g(x)=3$ $f(x-4)-2$, and describe the transformations to the graph of $f(x)$ needed to create the graph of $g(x)$.

Working from the inside out, $f(x-4)=(x-4)^2$, so the graph of g is a translation to the right of 4 units from the graph of f. $3f(x-4)$ stretches the graph by a factor of 3 from the x-axis. Finally, $3f(x-4)-2$ translates the graph down 2 units. Thus the graph of g is found by translating the graph of f right 4, stretching it from the x-axis by a factor of 3, and translating it down 2 units.

Exercises for Chapter 14

For Exercises 1–4, let $f(x)=\sqrt{x}$. For each of the equations given, describe the transformations needed to create the graph of the new function from the graph of $f(x)$, determine the domain and range of the new function, and make a pencil-and-paper sketch of the new function. Verify the graph with your graphing calculator.

1. $g(x)=f(x+2)-1$

2. $h(x)=-2f(x-3)+4$

3. $k(x) = f(-x) + 2$

4. $m(x) = 4f(-x+3) - 5$

For Exercises 5–8, let $f(x) = |x|$. For each of the equations given, describe the transformations needed to create the graph of the new function from the graph of $f(x)$, determine the domain and range of the new function, and make a pencil-and-paper sketch of the new function. Verify the graph with your graphing calculator.

5. $g(x) = -f(x-1) + 2$

6. $h(x) = 2f(x+1) - 3$

7. $k(x) = 1/2\, f(x) + 2$

8. $m(x) = 3f(-x-1) - 2$

For Exercises 9–12, let $f(x) = \dfrac{1}{x}$. For each of the equations given, describe the transformations needed to create the graph of the new function from the graph of $f(x)$, determine the domain and range of the new function, and make a pencil-and-paper sketch of the new function. Verify the graph with your graphing calculator.

9. $g(x) = f(x-1) + 2$

10. $h(x) = 2f(x+4) - 3$

11. $k(x) = -1/2\, f(x-3) + 2$

12. $m(x) = 3f(-x+1) - 4$

Determine the equations of the functions graphed in Exercises 13–16.

13.

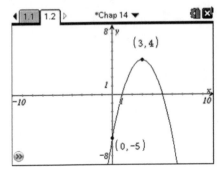

14.

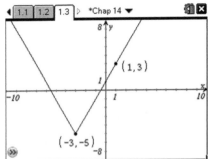

15.

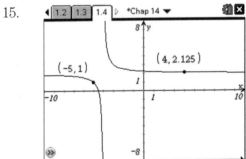

16.

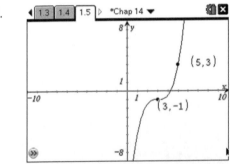

CHAPTER 15

Exponential Functions

Carbon-14 is a radioactive isotope of carbon and has a half-life of approximately 5,730 years. This means that half of a given mass of Carbon-14 will decay into a non-radioactive element in 5,730 years. 100 kg of Carbon-14 today will become 50 kg of Carbon-14 in 5,730 years (the rest is nonradioactive), then to 25 kg 11,460 years from now, then 12.5 kg in 17,190 years, etc. Carbon-14 was used to date the ages of artifacts found in archeological digs. The increased burning of fossil fuels since the beginning of the Industrial Revolution has increased the amount of Carbon-14 in the atmosphere so scientists have had to find another way to accurately measure the age of ancient objects.

Exponential versus Linear Growth

Sally is a new employee and is faced with a rather perplexing problem. Her new employer gave her a choice between two ways in which she could be paid each month. The first option has her being paid $100 on the first day of the month. Each day of the month after that, her pay will be $50 more than the previous day. The second option has her being paid $0.01 on the first day. Each day of the month after that, her pay will be twice as much as it was the day before.

Which option should she take?

The first salary option pays Sally $100 on the first day, pays her $150 on the second day, pays her $200 on the third day, and continues in such a way that on the last day of the month she will be paid $1,550. Her salary for the month will be $24,750. Very nice!

The second salary option pays her a penny on the first day, 2 cents on the second day, and 4 cents on the third day. This is not looking like a particularly good deal for the new employee. However, Sally is a patient person and continues to work out how much she will be paid for the month. The big smile on her face reveals that she will choose option 2. Sally will be paid $5,368,709.12 for the last day of the month and a total of $10,737,418.23 for the entire month!

When you examine the daily wage for each option, you see that the first option is linear, because the change in wage from day to day is always the same, $50. In other words, the daily change is found by adding. For the second option offered to Sally, however, the change in daily wage is found by multiplying, not adding. This is a classic example of an ***exponential function***, and exponential functions are the subject of this chapter.

Basic Exponential Functions: $f(x) = b^x$

The parent function for the exponential function is $y = b^x$, where b is a positive number not equal to 1. One is not included because 1 raised to any power is still 1, and the graph of this function is a horizontal line.

What are the domain and range of the exponential function? Because b is positive, the exponent can be any real number, so the domain is the set of reals. A positive number raised to any power will yield a positive answer.

(Recall that a negative exponent indicates a reciprocal, and the reciprocal of a positive number is a positive number. A fractional exponent indicates a radical, and the root of a positive number is a positive number.) Therefore, the range of an exponential function is $y > 0$. All exponential functions of the form $y = b^x$ pass through the point (0, 1).

The graph of an exponential function for values of $b > 1$ increases as the values of x increase. For example, look at the function $f(x) = 2^x$, its graph, and a table of values for this function.

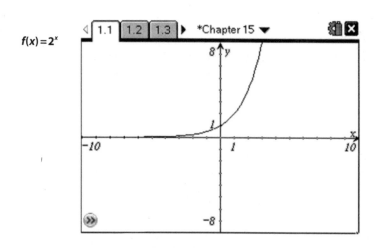

x	2^x
0	1
1	2
2	4
3	8
-1	1/2
-2	1/4
-3	1/8

As the values of x increase, the values of y grow continuously. As the value of x goes toward negative infinity, the value of y gets very close to 0. In fact, the line $y = 0$ becomes a boundary for the graph. The graph gets very close to this line but will not touch or cross it as these negative values of x become infinitely large in magnitude. This line is called a **horizontal asymptote**.

The rate at which the exponential graph will grow depends on the value of the base. For values larger than 2, the graph will grow that much more quickly; for values between 1 and 2, the graph will grow less quickly.

What happens if the value of b is between 0 and 1? Consider the function $g(x) = \left(\dfrac{1}{2}\right)^x$. When you examine the graph and its table of values, you see that $g(x)$ is a reflection of $f(x)$ across the y-axis.

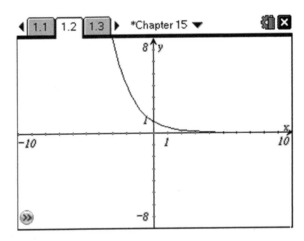

x	$\left(\dfrac{1}{2}\right)^x$
0	1
1	1/2
2	1/4
3	1/8
-1	2
-2	4
-3	8

The values of y decrease as the values of x increase, and the graph has a horizontal asymptote at $y = 0$.

Transforming Exponential Functions

It is very helpful to pay attention to three items when applying the rules of transformations to exponential functions. Be sure to focus on where the points $(0, 1)$ and $(1, b)$ move to and on what happens to the horizontal asymptote. Know these three things, you will be able to make a good estimate of the graph without a graphing calculator.

Describe the transformation of the parent exponential function that is necessary to graph $f(x) = 3\,(2)^{x-1} - 4$.

Because x is replaced by $x - 1$, the graph translates to the right 1. The leading coefficient of 3 causes the graph to be stretched from the x-axis by a factor of 3. Finally, the graph is translated down 4 units.

Translating $(0, 1)$ to the right 1 unit gives the point $(1, 1)$. Stretch this from the x-axis by a factor of 3 to get $(1, 3)$, and translate this point down 4 units to get the final image of $(1, -1)$.

Translating $(1, 2)$ to the right 1 unit gives the point $(2, 2)$. Stretch this from the x-axis by a factor of 3 to get $(2, 6)$, and translate this point down 4 units to get the final image of $(2, 2)$.

A horizontal translation and a stretch from the x-axis have no impact on the vertical asymptote. Translating the graph down 4 changes the horizontal asymptote to $y = -4$.

Describe the transformation of the parent exponential function that is necessary to graph $g(x) = -2\,(3)^{x+2} + 1$.

Because x is replaced by $x + 2$, the graph is translated to the left 2 units. The leading coefficient -2 causes the graph to be reflected over the x-axis and stretched from the x-axis by a factor of 2. Finally, the graph is translated up 1 unit.

$(0, 1)$ translates left to $(-2, 1)$. This point goes to $(-2, -2)$ after being reflected and stretched. It ends at $(-2, -1)$ after being translated up 1 unit.

$(1, 3)$ translates left to $(-1, 3)$, then to $(-1, -6)$, and finally to $(-1, -5)$.

The horizontal asymptote becomes $y = 1$.

An Application of Exponential Functions: Compound Interest

Simple interest, a topic you probably studied in grade 7 or 8, involves the computation of interest *once* during the completion of a financial transaction. The interest, I, is found by multiplying the amount of money borrowed, or the principal, P, the rate of interest, r, and the amount of time, t. The terms of the interest rate and the amount of time must agree in units. If the rate of interest is given in years, then the time must be given in years. For example, if $1000 is borrowed for 18 months at a rate of 6% per year, the amount of interest due is given by the equation $I = (1000)(0.06)(1.5) = \$90$. The time of 18 months is expressed as 1.5 years, and the interest rate is converted to a decimal for computational purposes.

ESSENTIAL

Most people have an understanding of borrowing money from a financial institution. Savings accounts and other investments are also loans in that the consumer is lending money to the institution. The big difference in these cases is the rate of interest given.

Compound interest is the mechanism by which most of the world's financial transactions take place. With compound interest, the interest is computed *periodically* throughout the term of the loan. In the example just used, suppose the $1,000 is borrowed for 18 months at 6% annual interest compounded quarterly. That is to say, the interest on the loan will be computed each quarter. The rate of interest is an annual 6%. Dividing this into four equal amounts (quarterly), the quarterly rate of interest (or periodic rate of interest) is 1.5%. To determine the amount of interest due at the end of the term of the loan, you will need to work through the process at each quarter. It is important to realize that the time unit in each of these calculations will be 1. The interest being computed is the interest for 1 quarter.

COMPOUND INTEREST

Quarter	Interest	Principal Due
1	$1800(0.015) = 27$	$1800 + 27 = 1827$
2	$1827(0.015) = 27.41$	$1827 + 27.41 = 1854.41$
3	$1854.41(0.015) = 27.82$	$1854.41 + 27.82 = 1882.23$
4	$1882.23(0.015) = 28.24$	$1882.23 + 28.24 = 1910.47$
5	$1910.47(0.015) = 28.66$	$1910.47 + 28.66 = 1939.03$
6	$1939.03(0.015) = 29.09$	$1939.03 + 29.09 = 1968.12$

The amount of interest due after 18 months (or 6 quarters of a year) is $168.12, much more than the $90 collected with simple interest. If you were doing the calculations, you may have noticed that the amount of interest due was not rounded by applying the usual rules for rounding but, rather, that the numbers were always rounded up to the next penny. The bank also keeps the fractional parts of a penny.

Look back on these calculations. The $1,827 due at the end of the first quarter was computed by $1800 + 1800(0.015)$. Factor the 1800 to get $1800(1.015)$. At the end of the second quarter, the $1800(1.015)$, or 1827, becomes $1800(1.015) + 1800(1.015)(0.015)$. Factoring the $1800(1.015)$, you get $1800(1.015)(1.015)$, or $1800(1.015)^2$. If you continue in this manner, the amount due at the end of the sixth quarter would be $1800(1.015)^6 = 1868.197$, or 1868.20. The difference of 8 cents are all those fractions of a penny that earned interest!

ESSENTIAL

The language of time divisions: annually—once per year; semiannually—twice per year; quarterly—4 times per year; monthly—12 times per year.

When P dollars are invested for t years at $100r$ % per year compounded n times per year, the value of the investment at the end of the term is given by $A = P\left(1 + \dfrac{r}{n}\right)^{nt}$.

$2000 is invested for 25 years at 3.4% compounded monthly. How much will the investment be worth at the end of 25 years?

$P=2000$, $r=0.034$, $n=12$, and $t=25$. The value of the investment is $A=2000\left(1+\dfrac{0.034}{12}\right)^{12*25}=\$4673.67$. (*Note:* Type the information into the calculator as it is written. Do not try to compute and enter the decimal for 0.034/12, and be sure that there are parentheses about 12*25.)

Which investment will be worth more at the end of 10 years, $1000 invested at 3.8% compounded semiannually or $1000 invested at 3.6% compounded monthly?

$P=1000$ and $t=10$ for both investments. The first investment has $r=0.038$ and $n=2$. It will be worth $A=1000\left(1+\dfrac{0.038}{2}\right)^{20}=1457.08$.

The second investment has $r=0.036$ and $n=12$. It will be worth: $A=1000\left(1+\dfrac{0.036}{12}\right)^{120}=1432.56$. The first investment will yield the larger amount.

Solving Exponential Equations

The basic premise behind solving exponential equations algebraically is that if the bases are equal, then the exponents must be equal.

Given $f(x)=5(2)^{x-3}-9$, solve $f(x)=151$.

Set $5(2)^{x-3}-9=151$. Add 9 and divide by 5 to get $(2)^{x-3}=32=2^5$. Set $x-3=5$ to find that $x=8$.

Solve: $128\left(\dfrac{2}{5}\right)^{x+4}+17=2017$

Subtract 17 and divide by 128 to get $\left(\dfrac{2}{5}\right)^{x+4}=\dfrac{125}{8}\bullet\dfrac{125}{8}=\left(\dfrac{5}{2}\right)^3=\left(\dfrac{2}{5}\right)^{-3}$. Setting $x+4=-3$ gives $x=-7$.

Solve: $4^x = 20$

20 is not a power of 4, nor are 4 and 20 powers of some common base. An algebraic solution is not likely to be possible. You can solve the problem graphically by graphing the functions $y = 4^x$ and $y = 20$.

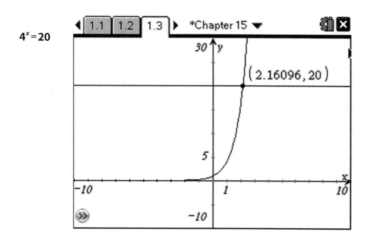

$4^x = 20$

The solution is approximately 2.16096.

e: A Natural Number

One property of compound interest that arises because it is an exponential function is that for a fixed rate of interest, the return is greater when the number of interest periods increases. That is, the return on a 4% loan will be greater when the money is compounded monthly than when it is compounded quarterly. This raises the question of whether there is a limit to the amount of return one can get. That is, the return on the 4% loan gets bigger as the number of interest periods grows without bound. The amount of principal does not affect the rate of return, so P is set equal to 1. To simplify the matter further, the rate of interest is set to 100%, and the time is set to 1 year. The formula for the amount of the investment becomes $A = \left(1 + \dfrac{1}{n}\right)^n$. As n grows without bound—that is, as n gets infinitely large—what happens to this value of A? Leonhard Euler (pronounced *oiler*) had done some work on this problem and found that the limit is approximately 2.71828. This number is called e in honor of Euler. Like π, e is a special number in mathematics, and just as you know that π is approximately 3.14159, you should know that e is approximately 2.711828. As you continue your studies in mathematics, you will see that many applications with exponential functions use e as a base.

Compute: e^2

TI-83/84: Type 2nd LN 2 to get 7.389056099
NSpire: Press e^x 2 to get 7.38906.

Solve: $e^x = 100$.

Graph the functions $y = e^x$ and $y = 100$ to find that $x = 4.60517$.

More Applications of Exponential Functions

Applications of exponential functions occur in the physical, life, and social sciences, as well as in the world of business.

A culture begins with 100 cells. Observation shows that each cell divides into 2 cells every 45 minutes. How many cells will be in the culture after 24 hours?

The initial amount, 100, will be the leading coefficient, because this is the number of cells present at time 0 (and $2^0 = 1$). The next thing to determine is the number of cell divisions that take place. The cells divide every 45 minutes, so you need to determine how many 45-minute periods exist in 24 hours; you find that there are $(24)(60)/45 = 32$. The number of cells present at the end of 24 hours will be $100(2)^{32} = 429{,}296{,}729{,}600$. (Using the analysis written, one can find the number of cells after m minutes by the formula $c(m) = 100(2)^{\frac{m}{45}}$.)

Data supplied by the Population Reference Bureau showed the United States population to be 307 million people in 2009 with a growth rate of

0.876% over the next 41 years. The equation modeling the U.S. population is $P(t)=307(1.00876)^t$, where t represents the number of years since 2009. On the basis of this model, predict the population of the United States in 2020.

2020 is 11 years after 2009, so the U.S. population for 2020 is predicted to be $P(11)=307(1.00876)^{11}=337.913$ million people.

Iodine-131 is a radioactive element used in medical treatments and has a half-life of 8.0197 days. If an initial dose of 10 milligrams (mg) is injected into a body, the amount of iodine-131 remaining after d days is given by the equation $I(d)=10\left(\dfrac{1}{2}\right)^{\frac{d}{8.0197}}$. How many milligrams of I-131 will remain in the body after 12 days?

$$I(12)=10\left(\frac{1}{2}\right)^{\frac{12}{8.0197}}=3.54458 \text{ mg}$$

Government regulations require banks to have a specified amount of cash available at all times. When a bank falls short of this amount, it borrows a sum of money from another institution for a short period of time. In such cases, the interest is computed continuously. Suppose Bank A needs to borrow \$2,500,000 from Bank B for 3 hours at a rate of 1.9% compounded continuously. The formula for the amount that the borrower will owe to the lender in such a transaction is $A=2500000e^{0.019t}$, where t is the amount of time in years. How much must Bank A pay in interest?

3 hours represents 1/8 of a day, or (1/8)(1/365) of a year. The amount of money Bank A must pay Bank B is $A=2500000e^{(0.019)(1/8)(1/365)}=2,500,016.27$.

Bank A will owe Bank B \$16.27 in interest.

Exercises for Chapter 15

In Exercises 1 and 2, for each function, describe the transformations applied to its parent function to arrive at the given function, and make a pencil-and-paper sketch of the given function. Verify your sketch with your graphing calculator.

1. $f(x) = 4(5)^{x+3} - 2$

2. $g(x) = -2(3)^{x-1} + 3$

3. $4,500 is invested for 5 years at 3.7% compounded quarterly. How much will the investment be worth at the end of this time?

4. Jackson has $6,000 to invest. In 5 years, he would like to have $10,000 to use as a down payment on a new car. If interest is compounded quarterly, at what rate must he invest his $6,000 in have $10,000 in 5 years? (A graphical solution is expected.)

5. Solve: $12(3)^{x+5} - 9 = 26235$

6. Solve: $960\left(\dfrac{3}{4}\right)^{x+2} + 129 = 356\dfrac{13}{16}$

7. Solve: $45(1.5)^{x-3} + 91 = 1283$

8. Estimates for the world population since 1990 are given by the equation $P(t) = 5344e^{0.012744t}$, where P is measured in millions of people and t is the number of years since 1990. What will the world's population be in 2015?

9. In living organic material, the ratio of the radioactive carbon-14 isotope present to the nonradioactive carbon-12 isotope present is about $1:10^{12}$. When the organic material dies, the nonradioactive isotope remains constant, but the radioactive isotope begins to decay with a half-life of 5,730 years. After the death of the organism, the ratio of the radioactive material present to the organic material present is given by $R = \dfrac{1}{10^{12}} e^{-\frac{t}{8267}}$. What will this ratio be for an organism that died 300 years ago?

Newton's Law of Cooling: Isaac Newton determined that the rate at which an object heats or cools is proportional to the difference between the temperature of the object and the temperature of the environment. Two important (but very different) applications for this principle can be found on the television shows *Emeril Live* (see Exercise 10) and *CSI* (see Exercise 11).

10. By the time a 14-pound turkey is taken out of the refrigerator, stuffed, and put into the 325°F oven, the temperature of the turkey is approximately 40°F. The temperature of the turkey t hours after it is put into the oven is given by $T(t) = 325 - 285e^{(-0.228x)}$. Dinner will be ready 30 minutes after the temperature of the turkey reaches 180°F. If you put the turkey into the oven at 1:00 o'clock, when do you eat?

11. The coroner has been summoned to the scene of a murder. The room in which the corpse was has a thermostat-controlled temperature of 70°F. Working from the assumption that the deceased had a body temperature of 98.6°F at the time of death, the coroner knows that the temperature of the deceased after t hours will be modeled by the equation $B(t) = 70 + 28.6e^{-0.1t}$. The coroner measures the temperature of the corpse and finds it to be 80°F. How long ago did the murder victim die?

CHAPTER 16

Logarithmic Functions

It is critical to be able to solve exponential equations that do not have a common base. For instance, how long will it take money deposited into an account that pays 6.8% interest compounded quarterly to double in value? A Scotsman by the name of John Napier devised such a process; he called it **logarithms**.

What Is a Logarithm?

The function $y=2^x$ is a one-to-one function; that is, it passes both the vertical-line test (therefore it is a function) and the horizontal-line test (therefore, its inverse is a function). The inverse of this function is $x=2^y$, which is found by interchanging the x- and y-coordinates. There is no algebraic operation that enables us to solve for y. This was Napier's creation. He determined that $x=2^y$ is the equivalent of $y=\log_2(x)$ (this is read, "y is the base-2 log of x"). When you compare the two equations, you recognize that y is the exponent of the equation $x=2^y$.

ESSENTIAL

A critical concept for you to remember is that a logarithm is an exponent. Too often people think of logarithms as something more than they are because they concentrate on memorizing formulas rather than understanding exactly what is happening.

Just as there are an infinite number of exponential functions with base b, there are an infinite number of logarithmic functions. The inverse of $y=b^x$ is $y=\log_b(x)$. The domain of the exponential function is the set of real numbers, and the range is the set $y > 0$. The domain of the logarithmic function is $x > 0$, and the range is the set of real numbers.

Determine the domain and range of $f(x)=\log_5(x)$.

The domain is $x > 0$, and the range is the set of real numbers.

Determine the domain and range of the function $g(x)=3\log_4(x+5)-2$.

The impact of $x+5$ is to shift the graph to the left 5 units, so the domain is $x > -5$. (If that is not clear to you, consider that $x+5 > 0$, so x must be greater than -5.) Multiplying the set of real numbers by 3 and then subtracting 2 does not change the set. The range is the set of real numbers.

Changing Exponential Equations to Logarithmic Equations

Learning to rewrite the form of exponential and logarithmic equations will enable you to solve many equations. If $c = b^a$, then the corresponding logarithmic equation is $a = \log_b(c)$.

Rewrite $8^2 = 64$ as a logarithmic function.

$c = 64$, $b = 8$, and $a = 2$. The corresponding logarithmic equation is $2 = \log_8(64)$.

Rewrite $16^{3/2} = 64$ as a logarithmic function.

$c = 64$, $b = 16$, and $a = 3/2$. The corresponding logarithmic equation is $3/2 = \log_{16}(64)$.

Rewrite $\log_{125}(25) = 2/3$ as an exponential equation.

$c = 25$, $b = 125$, and $a = 2/3$. The corresponding exponential equation is $125^{2/3} = 25$.

Rewrite $\log_{81}(27) = 3/4$ as an exponential equation.

$c = 27$, $b = 81$, and $a = 3/4$. The corresponding equation is $81^{3/4} = 27$.

Properties of Logarithms

There are a number of properties of exponential functions that translate into important logarithmic properties. Review the following table, where $A = b^m$ and $C = b^n$.

Exponential Rule	Logarithmic Rule	Verbal Translation
$AC = (b^m)(b^n) = b^{m+n}$	$\log_b(AC) = \log_b(A) + \log_b(C)$	The exponent for the product is the sum of the exponents.
$\dfrac{A}{C} = \dfrac{b^m}{b^n} = b^{m-n}$	$\log_b\left(\dfrac{A}{C}\right) = \log_b(A) - \log_b(C)$	The exponent for the quotient is the difference of the exponents.
$A^n = \left(b^m\right)^n = b^{mn}$	$\log_b\left(A^n\right) = n\log_b(A)$	The exponent for a term raised to a power is the product of the powers.

These properties will also be valuable in solving equations.

Simplify: $\log_b\left(\dfrac{x^2}{y^3 z^4}\right)$

The log of a quotient is the difference of the logarithms, so $\log_b\left(\dfrac{x^2}{y^3 z^4}\right)=$
$\log_b\left(x^2\right)-\log_b\left(y^3 z^4\right)$. The log of a product is the sum of the logarithms.

$$\log_b\left(x^2\right)-\log_b\left(y^3 z^4\right)=\log_b\left(x^2\right)-\left(\log_b\left(y^3\right)+\log_b\left(z^4\right)\right)$$

The exponent for a term raised to a power is a power.

$$\log_b\left(x^2\right)-\left(\log_b\left(y^3\right)+\log_b\left(z^4\right)\right)=2\log_b\left(x\right)-\left(3\log_b\left(y\right)+4\log_b\left(z\right)\right)$$

Distributing the minus sign, the final answer is:

$$\log_b\left(\dfrac{x^2}{y^3 z^4}\right)=2\log_b\left(x\right)-3\log_b\left(y\right)-4\log_b\left(z\right)$$

Solving Logarithmic Equations

Many logarithmic equations are solved by rewriting the equation as an exponential equation.

Solve: $\log_{16}(8)=a$

Rewriting the equation as an exponential equation, you get $8^a=16$. Getting a common base of 2, the equation becomes $\left(2^3\right)^a=2^4$, or $2^{3a}=2^4$. Setting the exponents equal yields $3a=4$, or $a=4/3$.

Evaluate: $\log_6\left(\sqrt[3]{36}\right)$

This is not an equation. Make it so by setting the expression equal to a. $\log_6\left(\sqrt[3]{36}\right)=a$ becomes $6^a=\sqrt[3]{36}=(36)^{1/3}$, which becomes $6^a=\left(6^2\right)^{1/3}=6^{2/3}$. Therefore, $a=2/3$.

Solve: $\log{(3x+4)}=2$

The log statement is written without a base, which indicates that this is the common (base-10) log. Writing this as an exponential equation, we have $3x+4=10^2=100$. Subtract 4 and divide by 3 to determine that $x=32$.

Solve: $\log_4(3x+4)=3$

The equation becomes $3x+4=4^3$, or $3x+4=64$. Subtract 4 and divide by 3 to get $x=20$.

Solve: $\log_b(81)=4$

The equation becomes $b^4=81$. Both 3 and -3 solve this equation, but you need to remember that the base must be a positive number. Therefore, $b=3$.

Solve: $\log_6(x-4)+\log_6(x+5)=2$

Exponents are added when terms are multiplied, so the left-hand side of the equation becomes $\log_6((x-4)(x+5))$. Rewriting the equation as an exponential equation yields $(x-4)(x+5)=6^2$. Solving the quadratic, you find that $x^2+x-20=36$ becomes $x^2+x-56=0$, and $(x+8)(x-7)=0$ yields $x=-8$ or 7. You can feel good about solving this equation, but be sure to check your answers. $x=-8$ causes the statement in the original problem to become $\log_6(-12)$, which is not defined. (Remember, the number inside the parentheses must be positive.) $x=7$ *does* work, so the answer is $x=7$.

Applications of Logarithmic Functions

The measure of sound intensity (decibels), the measure of acidity in a solution (the pH factor), and the measure of the extent to which the earth vibrates during an earthquake (the Richter value) are all examples of logarithmic functions.

pH measures the concentration of the hydrogen ion in a solution. (Brackets are used in chemical terminology to stand for concentration, so the symbol [H+] is read, "the concentration of the hydrogen ion.") pH is computed as the negative logarithm of [H+]; that is, pH = -log([H+]). A solution with a high concentration of the hydrogen ion—that is, a large [H+] value—is called an acid, and a solution with a low concentration of the hydrogen ion—that is, a low [H+] value—is called a base.

Determine the [H+] for lemon juice, which has a pH of 2.4.

pH = -log[H+] so for this problem, 2.4 = -log[H+]. Changing the equation to exponential form, we get [H+] = $10^{-2.4}$ = 0.003981.

The earth vibrates continuously but at different rates in different places. The normal intensity of the vibrations in the earth at a given location is designated by the variable I_0. The Richter scale compares the intensity of vibrations during a seismic event to this base number and reports the answer as a logarithm. That is, $R = \log\left(\dfrac{I}{I_0}\right)$, where I is the intensity of the vibration during the earthquake.

The earthquake that struck Port-au-Prince, Haiti, in 2010 measured 7.0 on the Richter scale. Determine the intensity of this quake in terms of the earth's normal vibrations in Port-au-Prince.

$R = \log\left(\dfrac{I}{I_0}\right)$ becomes $7 = \log\left(\dfrac{I}{I_0}\right)$. Written in exponential form, this becomes $10^7 = \dfrac{I}{I_0}$, or $I = 10^7 I_0$. The intensity of the vibrations during the earthquake in Port-au-Prince was 10 million times the normal vibrations of the earth at that location!

Logarithms are also used to solve exponential equations algebraically. One word of warning: Concentrate on solving the equation, and don't worry about the strange notation in the solution.

Similar to the measurement of vibrations of the earth during a seismic event, sound is measured relative to a normal value—in this case the threshold intensity of sound, I_0. The number of decibels, db, is given by $db = 10 \log\left(\dfrac{I}{I_0}\right)$.

Airport employees wear protective ear covers, because the decibel reading for planes on an airport runway is 120 db. How many times greater than the threshold of sound is the sound of a plane on a runway?

$120 = 10 \log\left(\dfrac{I}{I_0}\right)$ becomes $12 = \log\left(\dfrac{I}{I_0}\right)$. Rewrite as an exponential equation to get $10^{12} = \dfrac{I}{I_0}$. The intensity of sound on the runway, I, is 10^{12} times the threshold of sound, I_0.

Exercises for Chapter 16

In Exercises 1–4, determine the domain of each function.

1. $f(x) = \log_3(4x + 8)$

2. $g(x) = \log_7(4 - 3x)$

3. $p(x) = \log\left(\dfrac{x-1}{3}\right)$

4. $q(x) = \ln(x^2 + 1)$

In Exercises 5–8, rewrite each exponential equation in equivalent logarithmic form.

5. $7^3 = 343$

6. $8^{-2/3} = 1/4$

7. $25^{3/2} = 125$

8. $\left(\dfrac{2}{3}\right)^3 = \dfrac{8}{27}$

In Exercises 9–12, rewrite each of logarithmic equation in equivalent exponential form.

9. $\log_2(16) = 4$

10. $\log_9(27) = 3/2$

11. $\log_{100}(10) = 1/2$

12. $\log_{16}(1/8) = \text{-}3/4$

In Exercises 13–15, rewrite each logarithmic expression in terms of simple logarithmic expressions.

13. $\log_3\left((x+3)^2(x-1)\right)$

14. $\log\left(\dfrac{x^3 y^2}{z^8}\right)$

15. $\log_4\left(\dfrac{x^4}{\sqrt{y}}\right)$

Solve each equation in Exercises 16–19.

16. $8^{x-1}=16$

17. $5(6)^{2x-1}+37=217$

18. $3(4)^{x+3}-9=183$

19. $16\left(\dfrac{3}{4}\right)^x=\dfrac{81}{16}$

Solve each of the equations in Exercises 20–23. When necessary, round your answers to the nearest hundredth.

20. $5e^{2.3x}=80$

21. $23(4)^{0.1x}+17=148$

22. $18\left(\dfrac{3}{5}\right)^{x-2}+10=25$

23. $1200(1.034)^x=2000$

24. Find the hydrogen ion concentration, [H+], of a solution with pH 9.3.

25. The decibel level for busy street noise is approximately 67. How many times more than the threshold of sound is the sound on a busy street?

26. The earthquake that struck off the shore of Maule, Chile, in 2010 measured 8.8 on the Richter scale. Determine the intensity of this quake in terms of the earth's normal vibrations in Maule, Chile.

CHAPTER 17

Sequences and Series

Identifying patterns is a key piece of learning mathematics. When you were in elementary school, you were taught to "skip count"—to count by 2, count by 5, and count by 10. These processes, in addition to teaching you number sense, were the beginning of sequences and series.

Sequences: Number Patterns

Find the next three terms in each of the following examples.

1. 87, 89, 91, 93, ————, ————, ————

2. 128, 121, 114, 107, ————, ————, ————

3. 1250, 1330, 1410, 1490, ————, ————, ————

4. $\frac{1}{2}, \frac{2}{3}, \frac{3}{4}, \frac{4}{5}$, ————, ————, ————

5. $\frac{1}{3}, \frac{2}{5}, \frac{3}{7}, \frac{4}{9}$, ————, ————, ————

6. 1, 3, 6, 10, 15, ————, ————, ————

7. 1, 3, 9, 27, ————, ————, ————

RULE

A **sequence** is a set of numbers that represent the range of a function whose domain is the set of natural numbers.

Do you see the patterns? In Example 1, each term is 2 more than the previous term, so the next three terms are 95, 97, and 99. In Example 2, each term is 7 less than the previous terms, so the next three terms are 100, 93, and 86. The terms in Example 3 increase by 80, whereas the numerators and denominators each increase by 1 in Example 4. The numerators in Example 5 increase by 1, but the denominators increase by 2, so the next three numbers are 5/11, 6/13, and 7/15.

Example 6 is very different in that there is not a constant difference. The difference between the first two terms is 2, the difference between the second and third terms is 3, that between the third and fourth terms is 4, and that between the fourth and fifth terms is 5. The pattern in this sequence of numbers is that the difference between terms is increasing by 1; another way to express this is to say that the "second differences" $(3-2, 4-3,$ and $5-4)$ are constant. That means the next three terms are 21, 28, and 36.

Example 7 is also different in that the difference between terms is not constant, but in this case, neither are the second differences. The pattern in this example is to multiply the previous term by 3, so the next three terms are 81, 243, and 729.

Arithmetic Sequences

A sequence is **arithmetic** when the difference between terms is constant. The function defining an arithmetic sequence is linear.

Let $f(n)=3n+5$ define a sequence. Determine the first five terms of the sequence.

$f(1)=8, f(2)=11, f(3)=14, f(4)=17,$ and $f(5)=20$. The first five terms are 8, 11, 14, 17, 20.

Given the arithmetic sequence 2, 11, 20, 29, 38, . . . , determine the linear function that determines the sequence, and find the 40th term of the sequence.

The first term is 2 and the second term is 11. You can think of this as the ordered pairs (1, 2) and (2, 11). Knowing that the function is linear, compute the slope to be 9. A linear function takes the form $f(n)=mn+b$. Having computed the slope to be 9, you know that the function must be of the form $f(n)=9n+b$. $f(1)=2$ enables you to write the equation $2=9(1)+b$, so $b=$-7. The function that defines the sequence is $f(n)=9n-7$. The 40th term is $f(40)=9(40)-7=353$.

Find the 120th term of an arithmetic sequence in which the 10th term is 17 and the 32nd term is 193.

The tenth term is 17 ($f(10) = 10m + b = 17$), and the 32nd term is 193 ($f(32) = 32m + b$). Solve the system of equations

$10m + b = 17$

$32m + b = 193$

to find that $m = 8$ and $b = -63$. With $f(n) = 8n - 63$, $f(120) = 897$.

Arithmetic Series

A **series** is the sum of the terms in a sequence. If the sequence is arithmetic, then the corresponding series is naturally called an arithmetic series.

There is a famous story in the history of mathematics that took place at the end of the eighteenth century. The setting was a one-room school house in Germany. The students ranged from kindergarten age through eighth grade. The teacher, hoping to take a break for a few minutes, stood in front of the class and directed the students to find the sum of the first 100 counting numbers. Before the teacher could reach his desk, a small 4-year-old called out the answer. The method of solution was quite ingenious.

The student explained that he thought of the problem by writing the numbers forward and backward. In modern notation:

$1 + 2 + 3 + 4 + \cdots + 98 + 99 + 100$

$100 + 99 + 98 + 97 + \cdots + 3 + 2 + 1$

Adding vertically:

$101 + 101 + 101 + 101 + \cdots + 101 + 101 + 101$

This is the number 101 added 100 times, which the student knew to be 10,100. He also knew that this is twice the value of the true answer, so the sum of the first 100 counting numbers is 5050.

You might be thinking to yourself, "The kid was a genius." Well, he was. His name was Carl Friedrich Gauss. He grew up to be one of the most revered mathematicians and physicists in history.

Find the sum of the first 50 terms in the arithmetic sequence from Example 1.

You let S represent the sum of these numbers.

$S = 7 + 15 + 23 + \cdots + 391 + 399$

When you write the sum in reverse order, you get:

$S = 399 + 391 + 383 + \cdots + 15 + 7$

Now add these two equations together:

$2S = 406 + 406 + 406 + \cdots + 406 + 406$

$2S = 50(406)$

$S = (50/2)(406)$

$S = 10{,}150$

RULE

The sum, S, of the first n terms in an arithmetic series is $S = \dfrac{n}{2}(\text{first} + \text{last})$.

Find the sum of the first 80 terms of the arithmetic series whose terms are determined by the formula $f(n) = 9n - 7$.

The first term is $f(1) = 2$, and the 80th term is $f(80) = 713$. Therefore, the sum of the first 80 terms is $S = \dfrac{80}{2}(2 + 713) = 28{,}600$.

Find the sum of the first 90 terms in the arithmetic series $11+17+23+29+\cdots$.

The common difference between terms (which is also the slope of the defining function) is 6, so the defining function is of the form $f(n)=6n+b$. Use the first term, $f(1)=11$, to find $11=6(1)+b$, or $b=5$. The defining function is $f(n)=6n+5$. The 90th term of the sequence is $f(90)=545$. Thus, the sum of the first 90 terms is $S=\dfrac{90}{2}(11+545)=25{,}020$. The mathematical notation for this summation uses the uppercase Greek letter sigma, Σ (which represents summation), and reads $\displaystyle\sum_{n=1}^{90}6n+5$. The n is the variable of the problem, the 1 at the bottom of the notation indicates the first value of the domain, and the 90 above the sigma indicates the last value of the domain.

Geometric Sequences

A sequence is **geometric** when the ratio between successive terms is a constant. The function that defines the geometric sequence is exponential.

Find the first 5 terms and the 20th term of the geometric sequence generated by the function $f(n)=\dfrac{2}{27}(3)^n$.

The first term is $f(1)=2/9$. The second through fifth terms are 2/3, 2, 6, 18, and 54. The 20th term is $f(20)=258{,}280{,}326$. Yes, the values of a geometric sequence get very large very quickly.

Given the geometric sequence 1600, 800, 400, 200, 100$\cdots$, determine the exponential function that determines the sequence, and find the 25th term of the sequence.

The basic form of an exponential function is $f(n)=a(r)^n$. Here, $f(1)=1600=ar$, and $f(2)=800=ar^2$. Solve the following systems:

$$800 = ar^2$$

$$1600 = ar$$

by dividing to determine that $b = 1/2$. Solve and find that $a = 3200$. The exponential function that generates the geometric sequence is $f(n) = 3200\left(\dfrac{1}{2}\right)^n$, or $f(n) = 3200(2)^{-n}$. The 25th term in the sequence is

$$3200(2)^{-25} = \frac{25}{262144}.$$

ESSENTIAL

The common factor in a geometric sequence is also the base of the exponential function that defines the sequence.

Find the 30th term of a geometric sequence in which the 5th term is 18 and the 12th term is 294,912.

$$f(5) = 18 = ar^5 \text{ and } f(12) = 294{,}912 = ar^{12}$$

$$294{,}912 = ar^{12}$$

$$18 = ar^5$$

Solve the system by dividing to get $16384 = r^7$, or $r = 4$. Solving for a, determine that $a = \dfrac{9}{8192}$. The 30th term of the sequence is 1,266,637,395,197,952.

Find the next three terms of the sequence: 1, -2, 4, -8, _____, _____, _____

The constant factor for this sequence is -2, so the next three terms will be 16, -32, and 64. A complication arises in writing the exponential function that generates this sequence. Technically speaking, the base of an exponential function must be a positive number (but not 1). This is because values such as $(2)^{-1/2}$ are not real numbers, so the graph of the exponential function would have a lot of problems associated with it.

Because the domain of the function defining a geometric sequence is the set of natural numbers, there are no input values that create trouble-some situations. Therefore, rather than writing a function such as $f(n) = (-1)^n (-1/2)(2)^n$ to define this geometric sequence, it will be easier for you to create this sequence by writing $f(n) = (-1/2)(-2)^n$.

Geometric Series

The sum of the terms in a geometric sequence is called a **geometric series**. The process for computing the sum of the terms in a geometric series is different from the method for solving the arithmetic series.

Find the sum: $1+2+4+8+16+\cdots+2048$

Let S represent the sum of the series.

$S = 1+2+4+8+16+\cdots+2048$

Multiply both sides of the equation by the common ratio. It is con-venient to offset the product (as shown here).

$S = 1+2+4+8+16+\cdots+2048$

$2S = \quad 2+4+8+16+\cdots+2048+4096$

Subtract the two equations. (Observe how the middle terms drop out of the problem.)

$-S = 1-4096$

$S = 4095$

The function defining the terms in this series is $f(n) = \dfrac{1}{2}(2)^n$. The for-mula for finding the sum is:

$$S = \frac{a_1\left(1-r^n\right)}{1-r}, \text{ where } a_1 \text{ represents the first term in the series.}$$

Using summation notation, the sum can be computed as $\sum_{n=1}^{12}\left(\frac{1}{2}(2)^n\right)$.

Find the sum of the first 40 terms of the geometric series whose terms are determined by the formula $f(n)=1200\left(\frac{3}{4}\right)^n$.

Using the summation notation yields $S=\sum_{n=1}^{40}1200\left(\frac{3}{4}\right)^n=3599.96.$ By this formula, $S=\dfrac{900\left(1-\left(\frac{3}{4}\right)^{40}\right)}{1-\dfrac{3}{4}}=3599.96.$

Find the sum of the first 25 terms in the geometric series $4+12+36+108+\cdots$.

The common ratio is 3. $f(1)=4=a(3)$ implies that $a=4/3$. The function that generates the terms of the series is $f(n)=\dfrac{4}{3}(3)^n$. Using summation notation, you find that the sum is $S=\sum_{n=1}^{25}\dfrac{4}{3}\left((3)^n\right)$, and by this formula, $S=\dfrac{4(1-3^{25})}{1-3}=169{,}457{,}721{,}884.$

You notice that this is a very large number, and although the previous example had more terms, the sum was very close to 3600. The base of $f(n)=1200\left(\frac{3}{4}\right)^n$ is a number smaller than 1. The larger the domain value applied to this function, the smaller the term gets, and the less it adds to the accumulated sum. In fact, if the absolute value of the base of the exponential function is less than 1, the sum of an infinite number of terms will reach a finite value.

RULE

The sum of an infinite geometric series is $S=\dfrac{a_1}{1-r}$ when $|r|<1$.

Find the sum: $2000 + 1000 + 500 + 250 + \cdots$

The first term of the series is 2000, and the common ratio is 1/2. Therefore, the sum is: $S = \dfrac{2000}{1 - \dfrac{1}{2}} = 4000.$

Find the value of $\displaystyle\sum_{n=1}^{\infty} (-18)\left(\dfrac{-2}{3}\right)^n .$

The first term is $f(1) = (-18)(-2/3) = 12$, and the common ratio is -2/3. Therefore, the sum of the infinite number of terms for this series is $S = \dfrac{12}{1 - \left(\dfrac{-2}{3}\right)} = \dfrac{12}{\dfrac{5}{3}} = \dfrac{36}{5}.$

An Application of Geometric Series: Annuities

Saving money for your future is an important—and wise—course of action to take. One popular form of saving is called an individual retirement account (IRA). You deposit money t into an account (tax-free), and t accumulates interest. You do not have to pay taxes on this money until you withdraw it (which is usually at age 59.5 or later, when you might retire. This summary is a simplified version of the entire process, and you should do some research to learn more about it.

An annuity is a financial situation in which the timing of the payments coincides with an interest period. Loan payments are a type of annuity in that the lump sum is given to the borrower "up front" and is repaid in equal periodic installments over a period of time. For instance, traditional car and mortgage payments are repaid monthly, and many college loans are repaid quarterly. A **simple annuity** is one in which equal payments occur in equal time frames. The process of determining the amount of each payment is called **amortizing**.

As an example, consider the savings plan of depositing $2000 into an account that pays 3.8% interest compounded annually. You make the first deposit on your 25th birthday and the last deposit on your 62nd birthday. How much money will be in the account after your last deposit?

The first deposit will collect interest for 37 years and will grow to $2000(1.038)^{37}$ dollars. The deposit made on your 26th birthday will grow to $2000(1.038)^{36}$ dollars.

ESSENTIAL

Many advanced calculators have a built-in financial function that can perform these computations for you without your having to create a series equation.

The amount of money in the account after all the deposits have been made will be:

$$2000(1.038)^{37}+2000(1.038)^{36}+2000(1.038)^{35}+\cdots+2000(1.038)+2000$$

For ease of computation, write these numbers in reverse order:

$$2000+2000(1.038)+\cdots+2000(1.038)^{35}+2000(1.038)^{36}+2000(1.038)^{37}$$

The first term is 2000, the common ratio is 1.038, and the last exponent is 37. The sum of money in the account after the last deposit is

$$S=\frac{2000\left(1-(1.038)^{37}\right)}{1-1.038}=156,562.06.$$

You make 38 deposits of $2,000, for a total of $76,000, and collect more than $80,000 in interest. This is the power of compound interest!

Exercises for Chapter 17

For Exercises 1–5, find the next three terms for each of the following sequences.

1. 11, 30, 49, 68, …

2. 24, 18, 12, 6, …

3. 24000, 18000, 13500, 10125, …

4. 5, -15, 45, -135, …

5. 1, 4, 9, 16, …

6. Find the first 3 terms of the arithmetic sequence defined by
   $f(n) = 12n + 9$.

7. Find the first 3 terms of the arithmetic sequence defined by
   $g(n) = -7n + 85$.

8. The 12th term in an arithmetic sequence is 57, and the 19th term is 92. Find the 40th term of this sequence.

9. Find the sum of the first 60 terms of the arithmetic series
   $5 + 9 + 13 + 17 + \ldots$

10. The 3rd term of an arithmetic sequence is 19, and the 18th term is 124. Find the sum of the first 20 terms of this sequence.

11. Find: $\sum_{n=1}^{35} 9n + 11$

12. Find the first three terms of the geometric sequence defined by
    $f(n) = 3(4)^n$.

13. Find the first three terms of the geometric sequence defined by
    $g(n) = 2(-3)^n$.

14. Find the first three terms of the geometric sequence defined by $k(n) = 1{,}000{,}000(0.1)^n$.

15. The 4th term of a geometric sequence 54, and the 7th term is 1458. Find the 11th term.

16. Find the sum of the first 20 terms of the geometric series $1 + 2 + 4 + 8 + \cdots$.

17. Find the sum of the first 30 terms of the geometric series $8100 + 2700 + 900 + 300 + \cdots$.

18. Find: $\displaystyle\sum_{n=1}^{10} 8(3)^n$

19. Find: $\displaystyle\sum_{n=1}^{\infty} 30\left(\frac{2}{3}\right)^n$

20. Brendon and Stacey have set up a college fund for their newborn son. They make the first deposit the day he was born and continue to make annual deposits on his birthday until he is 18 years old. If each deposit is $5,000 and the account pays 5% compounded annually, how much money will be in the account after the deposit is made on his 18th birthday?

Regression with Graphing Calculators

Creating models for sets of data is a key skill for the businessperson, scientist, politician, or anyone else who uses data to make decisions. This is because modeling enables one to make predictions. The graphing calculator can help with this task. Given a set of data, one can create a model that is a linear, quadratic, cubic, quartic, exponential, logarithmic, sinusoidal, logistical, or power function. Determining which model is the most appropriate model is addressed in statistics courses, although a fundamental rule of thumb has to do with the plot of the residuals of the regression.

This appendix shows how to use the two most popular graphing calculators—the TI 8xx series (83+, 84, and 84+) and the TI Nspire—in working through three problems: a linear regression based on data obtained from a group of students reading a tongue twister, a quadratic regression based on the heights of a bouncing ball measured with a motion detector, and an exponential regression using the maximum heights to which the ball bounces. The keystrokes and screen shots are given so that you can use the tool.

TI 8xx Series

Five volunteers are asked to read this tongue twister: How many boards could the Mongols hoard if the Mongol horde got bored?

The volunteers are numbered 1 through 5. Volunteer 1 reads the tongue twister, and the time needed to read the tongue twister, as measured with a stopwatch, is recorded. The watch is reset, and volunteer 1 again reads the tongue twister, with the difference that volunteer 2 will begin reading the tongue twister when volunteer 1 has finished. When the second person has finished reading, the data are recorded on the second line. This process is extended to the third volunteer, then to the fourth, and finally to the fifth. The data in the table represent the number of people who read (column L1) and the time, in seconds (column L2).

Press STAT, option 1: Edit and enter the data.

Data for Tongue Twister

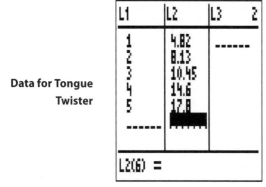

Before creating the scatter plot for the data, go to the equation editor (Y=) and either turn it off (put the cursor on the equals sign, and press ENTER so that the equals sign is no longer back-lit) or clear any active equations. Press 2nd Y= to go to the STAT PLOT menu. Press ENTER for Plot 1. With the cursor on the ON option, press ENTER.

STAT Plot

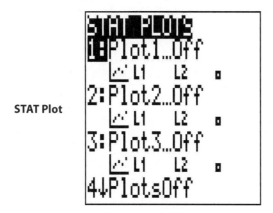

Press the ZOOM button, and choose option 9:ZSTAT to have the calculator determine the window for the scatter plot. You can always adjust the window through the WINDOW button.

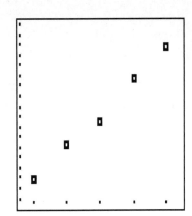

It looks like a line will pass through these points. A linear regression is a good option.

The calculator contains a diagnostic command that will return the values of the correlation coefficient, r, and the coefficient of determination, r^2. Turn this on by pressing 2nd 0 to go to the Catalog, x^{-1} to go to the D's. Scroll down until you reach the Diagnostic On command.

Diagnostic On

```
CATALOG           ▣
 DelVar
 DependAsk
 DependAuto
 det(
 DiagnosticOff
 DiagnosticOn
▶dim(
```

Have the calculator determine the line of best fit by pressing the STAT button, right arrow to the CALC menu, and choose option 4: LinReg(ax+b).

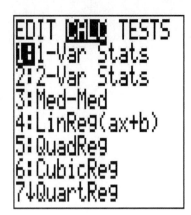

STAT CALC Menu

The calculator requires two pieces of information for the LinReg command: the list containing the input values and the list containing the output values. You have the option of including the location where you would like the equation stored. Before proceeding, take a moment to look at your calculator. Above the numbers 1–6 you will find L1–L6 on the calculator plate. These are the six default lists that come with the calculator.

1. Press 2nd 1 to select L1 as the input list.
2. The input list and output list must be separated by a comma, which you will find on the row above the 7.
3. Press 2nd 2 to select L2 as the output values.
4. Put a comma after L2 and enter the location of the equation in which you would like to store the data.
5. To the left of the CLEAR button is the VARS button. Press this button and the right arrow to activate the Y-VARS menu. Choose option 1:Functions. You have access to the ten Y1–Y0 that come with the calculator. Choose one of these.

Press ENTER to complete the regression.

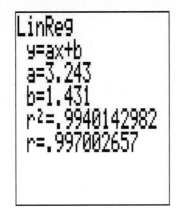

Press GRAPH to see how well this equation fits the data.

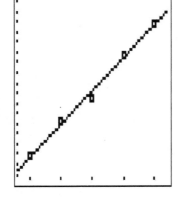

The slope of the regression line is 3.243. The number is interpreted to mean that it takes an average of 3.243 seconds for the tongue twister to be read. The value of the vertical intercept—the time needed for 0 people to read the tongue twister—has no pertinent meaning in this case.

Ball Bounce

A ball is released in free fall, and its height is measured by a motion detector over a 3-second period. The time and height (in meters) are stored into the graphing calculator. A sample of the data points is shown, as well as the scatter plot for all the points.

L1	L2	L3	1
0	0	------	
.04301	.00425		
.08602	1.2753		
.12902	1.24		
.17203	1.1545		
.21504	1.0524		
.25805	.92413		

L1(1)=0

Data from Ball Bounce

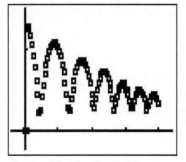

Scatter Plot for the Ball Bounce

The scatter plot shows a series of parabolic arches. The TRACE command was used to find the two lowest points and the highest point for one complete bounce, and a quadratic regression [STAT CALC 5:QUADREG] was performed for these data.

L3	L4	L5	5
.4731	.2416	------	
.9031	1.0725		
1.3332	.2423		
------	------		

L5(1)=

Data from One Arc of the Ball Bounce

QuadReg
y=ax²+bx+c
a=-4.490843345
b=8.112624192
c=-2.591325736
R²=1

Quadratic Regression for the Ball Bounce

Even with the relatively crude equipment with which the measurements were taken, the leading coefficient -4.49 is very close to the theoretical gravitational coefficient of -4.9 meters per second per second (m/sec/sec).

Exponential Regression

The TRACE command was used to determine the highest points of the ball on each bounce of the experiment done with the ball bounce. The data and scatter plot are shown.

Maximum Heights of Ball Bounce

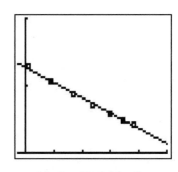

Scatter Plot for the Maximum Heights

The scatter plot looks like it might represent linear data. The regression equation and graph of this model are as follows:

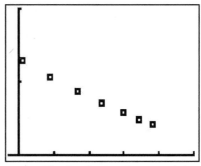

Linear Reg Maximum Heights

Lin Reg Model for the Maximum Heights

The regression equation [STAT CALC 0:EXPREG] and graph of the model for the exponential model show a much better fit.

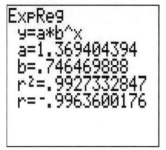

Exponential Reg Maximum Heights

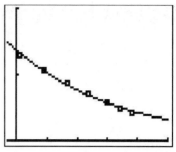

Exp Reg Model for the Maximum

The base of the exponential function, 0.746, represents the ratio of the heights of the ball from one bounce to the next. That is, the ball bounces to almost 75% of its previous height.

TI Nspire

Five volunteers are asked to read this tongue twister: How many boards could the Mongols hoard if the Mongol horde got bored?

The volunteers are numbered 1 through 5. Volunteer 1 reads the tongue twister, and the time needed to read the tongue twister, as measured with a stopwatch, is recorded. The watch is reset, and volunteer 1 again reads the tongue twister, with the difference being that volunteer 2 will begin reading the tongue twister when volunteer number 1 has finished. When the second volunteer is finished reading, the data are recorded on the second line. This process is extended to the third volunteer, then to the fourth, and finally to the fifth. The data in the table represent the number of people who read in column A and the time, in seconds, in column B.

Data for
Tongue Twister

Insert a new graph page (Control I). From the Menu, 3:Graph Type, 4: Scatter Plot. Enter Readers for the x-values and Time for the y-values. Press ENTER.

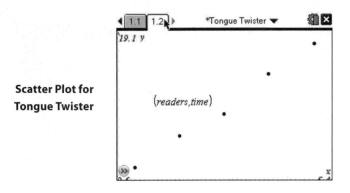

Scatter Plot for Tongue Twister

To determine the equation of the line of best fit, press CONTROL left arrow to go back to the spreadsheet, and put the cursor in cell C1. (You can choose to put the cursor in any of the columns, but C is chosen for ease.) Press MENU, 4:Statistics, 1: Stat Calculations, 3:Linear Regression (mx+b). [Option 4:Linear Regression (a+bx) is traditionally used in college statistics courses because the use of subscripted variables for multivariable regressions takes the form $y = b_0 + b_1 x_1 + b_2 x_2 + \cdots$]

Linear Regression on the Nspire

A window appears with a number of drop menus. Use the TAB key to move from one drop menu to the next. In the first menu, press the down arrow and choose Readers as the input values and Time for the output values in the next menu. The default location for the regression equation is the first free function available. Move to the bottom of the list to choose column c[] for the first result column.

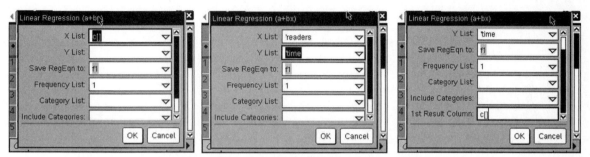

Entering the Location of the Data for the Regression

The results of the regression are displayed in columns C and D on the spreadsheet.

Results of the Linear Regression

Go back to the graph page (CONTROL right arrow). MENU, 3:Graph Type, 1:Function. The function f2(x) will appear on the bottom of the page. Up arrow once to f1(x), where you will see the regression equation. Press ENTER to have the equation graphed.

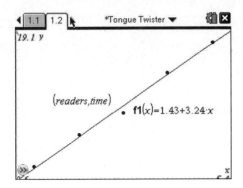

Scatter Plot and Model for the Tongue Twister

Ball Bounce

A ball is released in free fall, and its height is measured by a motion detector over a 3-second period. The time and height (in meters) are stored into the graphing calculator. A sample of the data points is shown, as well as the scatter plot for all the points.

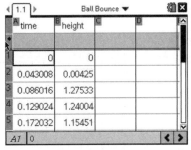

Data from Ball Bounce

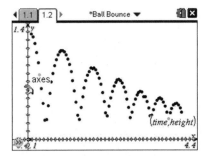

Scatter Plot for the Ball Bounce

The scatter plot shows a series of parabolic arches. The TRACE command was used to find the two lowest points and the highest point for one complete bounce, and a quadratic regression [MENU, 4:STATISTICS, STAT CALCULATIONS 6:Quadratic Regression] was performed for these data.

Quadratic Regression for One Arc of the Ball Bounce

Even with the relatively crude equipment with which the measurements were taken, the leading coefficient -4.49 is very close to the theoretical gravitational coefficient of -4.9 meters per second per second (m/sec/sec).

Exponential Regression

The TRACE command was used to determine the highest points of the ball on each bounce of the experiment done with the ball bounce. The data and scatter plot are shown.

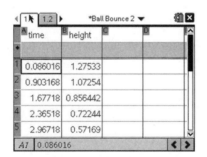

Maximum Heights of Ball Bounce

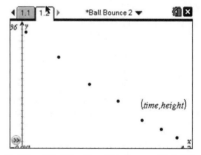

Scatter Plot for the Maximum Heights

The scatter plot looks like it might represent linear data. The regression equation and graph of this model are as follows:

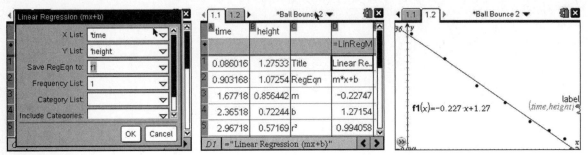

Linear Regression Model for the Maximum Heights

The regression equation [MENU, 4 STATISTICS, STAT CALCULATIONS A:Exponential Regression] and graph of the model for the exponential model show a much better fit.

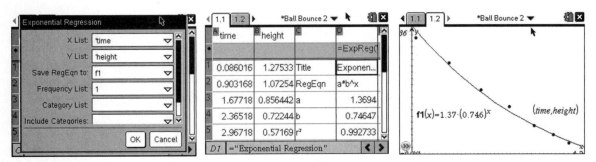

Exponential Regression Model for the Maximum Heights

The base of the exponential function, 0.746, represents the ratio of the heights of the ball from one bounce to the next. That is, the ball bounces to almost 75% of its previous height.

Glossary

Symbols

>
Is greater than

≥
Is greater than or equal to

<
Is less than

≤
Is less than or equal to

Algebra Terms

Abscissa
The first number in an ordered pair.

Additive Identity
$n+0=0+n=n$

Additive Inverse
$n+(-n)=0$

Arithmetic sequence
A sequence in which the difference between terms is a constant.

Arithmetic series

The sum of the terms in an arithmetic sequence.

Associative Property of Addition

$(a+b)+c=a+(b+c)$

Associative Property of Multiplication

$(ab)c=a(bc)$

Break-even point

The value for the number of items sold such that revenue $=$ cost.

Commutative Property of Addition

$a+b=b+a$

Commutative Property of Multiplication

$ab=ba$

Conjugates

Numbers that come in the pairs $a+b$ and $a-b$

Demand function

An equation that relates the number of items that consumers are willing to purchase and the price per item.

Discriminant

b^2-4ac, the number found within the square root of the quadratic formula.

Distributive Property of Multiplication over Addition

$a(b+c)=ab+ac$

Equilibrium point

The point at which the supply function $=$ the demand function. The price and number of items at which all that the suppliers put on the market will be purchased.

Geometric sequence
A sequence in which the ratio of successive terms is a constant.

Geometric series
The sum of the terms in a geometric sequence.

Index
The number that corresponds to the power in the power function. The n, in $a = \sqrt[n]{b}$ corresponds to $b = a^n$.

Integer
$\ldots, -4, -3, -2, -1, 0, 1, 2, 3, 4, \ldots$

Intercepts
The points at which a graph intersect the coordinate axes.

Irrational number
Any number that cannot be written as the ratio of two integers.

Multiplicative Identity
$n(1) = (1)n = n$

Multiplicative Inverse
$n\left(\dfrac{1}{n}\right) = \left(\dfrac{1}{n}\right)n = 1$

Natural number (counting number)
$1, 2, 3, 4, 5, \ldots$

Ordered Pair
The location (a, b), where a is the horizontal position from the origin and b is the vertical position from the origin.

Ordinate
The second number in an ordered pair.

Origin

The intersection of the x- and y-axes, which has coordinates $(0, 0)$.

Profit

The difference between revenue and cost.

Rational number

Any number that can be written as the ratio of two integers. Any number that can be written as a fraction, as a terminating decimal, or as a repeating decimal.

Radicand

The number under the radical.

Revenue

The income earned from selling items.

Roots

The value(s) of the variable that solve(s) an equation.

Sequence

The output of a function whose domain is the set of natural numbers.

Series

The sum of the numbers in a sequence.

Supply function

The equation that relates the number of items that suppliers are willing to put into a market at a given price per item.

Whole number

$0, 1, 2, 3, 4, \ldots$

Zeroes of a Function

Values of the input variable that cause the output to be zero.

Answers to Exercises

Chapter 1

1. N, W, Z, Q, R

2. Z, Q, R

3. Q, R

4. I, R

5. -19

6. 1

7. 90

8. -15

9. 15

10. -4

11. -4

12. 4

13. -2

14. 18

15. 10

16. 25

17. 27

18. -8

19. 50

20. 1

21. -12

22. 16

23. -125

24. -1

25. 66

26. 3

27. $3x$

28. $-6y + 13w$

29. $5x - 2x^2$

30. $-7y + 10y^2 - 2y^3$

31. $18f - g$

32. $12q + 35r$

33. $63x^2$

34. $20y^3$

35. $216x^6$

36. $72p^5$

37. $\dfrac{3m^2}{2}$

38. $\dfrac{2h^2}{3k^4}$

39. 1

40. $\dfrac{64v^3}{27}$

Chapter 2

1. 14

2. $\dfrac{301}{15} = 20\dfrac{1}{15}$

3. 2.34

4. -62

5. $5/9$

6. 35.1

7. 9

8. 6

9. 13.5

10. 10

11. $\dfrac{122}{15} = 8\dfrac{2}{15}$

12. -11

13. 30

14. $\dfrac{-111}{16} = -6\dfrac{15}{16}$

15. 75

16. 27

17. -29

18. $g < 6.25$

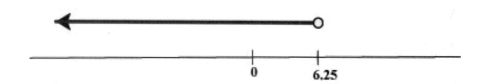

19. $r \geq -5$

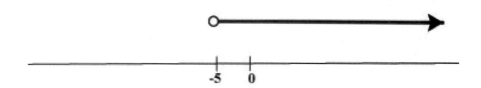

20. $n > -5$

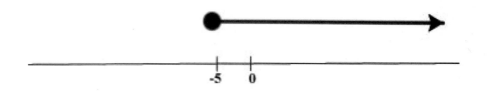

21. $-9 < x \leq 11$

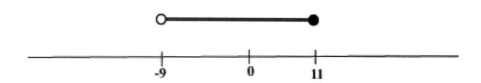

22. -6 > x > -18

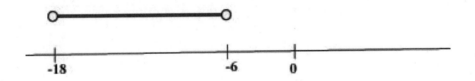

23. $x < 7$ or $x > 22$

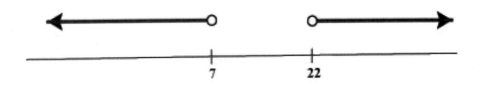

24. All real numbers

25. 80 quarters, 180 dimes, 210 quarters (*Hint:* Let q=number of quarters.)

26. 81.25 mph (In the 2 hours that Alison will drive, Sonya will have driven at most 162.5 miles.)

27. Amy is eighteen and thus is eligible to vote in the United States.

28. 33 pieces

Chapter 3

1. $22x^2 - 3x - 1$

2. $8x^2 - 21x + 7$

3. $2x^3 - 3x^2 - 4x + 21$

4. $24x^2 + 5x - 14$

5. $6w^2 - 23wp - 18p^2$

6. $18x^2 - 39x + 20$

7. $15w^2 + 31wv + 14v^2$

8. $49x^2 - 4$

9. $\dfrac{4}{9}r^2 - \dfrac{25}{49}$

10. $108x^3 - 159x^2 + 11x + 40$

11. $16c^3 - 18c^2d - 19cd^2 + 15d^3$

12. $16m^2 + 56m + 49$

13. $9k^2 - 24kz + 16z^2$

14. $64m^3 + 336m^2 + 588m + 343$

15. $27k^3 - 108k^2z + 144kz^2 - 64z^3$

16. $512g^3 - 27$

17. $1000r^3 + 729q^3$

18. $4x - 5$

19. $8z - 5$

20. $3q + 2 + \dfrac{1}{5q + 3}$

21. $4w^2 + 1 - \dfrac{5}{7w + 8}$

22. $4w^2 - 9$

23. $3d^3 + 5d^2 + 7$

24. $P = 24x + 24; A = 27x^2 + 66x + 35$

Chapter 4

7. A: Quadrant II; B: Quadrant IV; C: axis; D: Quadrant I; E: Quadrant III; F: axis

8. x-int: 5/6; y-int: -5; slope: 6

9. x-int: -28/3; y-int: 7; slope: 3/4

10. x-int: 6; y-int: -8; slope: 3/4

11. x-int: -1; y-int: -2/3; slope: -2/3

12. x-int: 25/8; y-int: -25/3; slope: 8/3

13. $y = 4x + 13$

14. $y = -7x + 21$

15. $y = 4/5x + 11$

16. $y = 3x + 2$

17. $y - 7 = -2/3\ (x - 4)$

18. $y - 4 = 9/7\ (x + 1)$

19. $y + 5 = -12/5\ (x - 3)$ or $y - 7 = -12/5\ (x + 2)$

20. $y = 7/16\ (x + 5)$ or $y - 7 = 7/16\ (x - 11)$

21. $12x + 5y = 11$

22. $-7x + 16y = 35$

23. $(8, 0)\ ;\ (0, -18)$

24. $9/4$

25.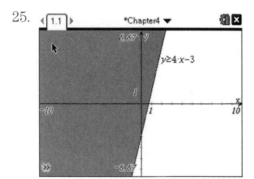

26.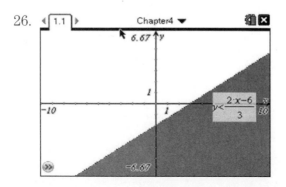

Chapter 5

1. (a) D:{-2, 0, 1, 3, 5}; R:{1, 2, 4, 5}

   (b) Function

   (c) A^{-1} = {(4, 3), (2, 5), (1, 0), (4, 1), (5, -2)}

   (d) Not a function

2. (a) D:{-1, 3, 5}; R:{1, 2, 3, 5}

   (b) Not a function

   (c) B^{-1} = {(2, 5), (3, -1), (5, 3), (1, 5)}

   (d) Function

3. 20

4. 1

5. 3.5

6. 8/7

7. 20

8. -14

9. 1

10. -12

11. 17

12. 31

13. 7

14. 5.5

15. $\dfrac{3(3x+8)-4}{2(3x+8)-1}=\dfrac{9x+20}{6x+15}$

16. $x+3$

17. $2\left(\dfrac{3x-4}{2x-1}\right)-3=\dfrac{2(3x-4)-3(2x-1)}{2x-1}=\dfrac{-5}{2x-1}$

18. $f^{-1}(x)=\dfrac{x-8}{3}$

19. $k^{-1}(x)=2(x-3)$

20. $x\neq 1/2$

Chapter 6

1. (-4, 5)

2. (-5, 3)

3. (-5, 5)

4. (3, -3)

5. (3/8, -3/4)

6. (4.7, -2.3)

7. (-1.64, 3.28)

8. (9/7, -5/3)

9. (-142, 107)

10. (-4, -3, 11)

11. (-5, 8)

12. (-4, -3, 11)

13. (1/2, -2/3, 7/12)

14. Sunny had 48 dimes and 58 quarters.

15. There are 215 females and 209 males in the class.

16. Mrs. Stockwell needs to replace fifteen $75 books and eight $25 books.

17. Tim has 77 nickels, 233 dimes, and 140 quarters.

18. The speed of the plane in still air is 225 mph, and the speed of the wind is 25 mph.

19. Point in shaded region

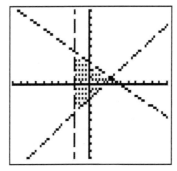

20. Point in shaded region

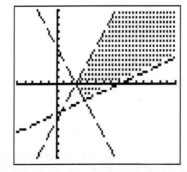

Chapter 7

1. $8x(x-3)$

2. $4x(2x-3)$

3. $4x^2(2x^2-3)$

4. $(x-11)(x+11)$

5. $(9x-5)(9x+5)$

6. $9(3x-5)(3x+5)$

7. $(4p-5)^2$

8. $(8r+7)^2$

9. $2m\,(3m+5)^2$

10. $(5t+4)(25t^2-20t+16)$

11. $(3-2q)(9+6q+4q^2)$

12. $(3x-2)(3x+2)(9x^2+6x+4)(9x^2-6x+4)$

13. $(x-7)(x-5)$

14. $(g+4)(g+15)$

15. $(8c+5)(2c-9)$

16. $(4x+9)(9x+5)$

17. $(4q-5)(5q-4)$

18. $(4q-5)(5q+4)$

19. $2m(m^2-m-35)$

20. $(5x+4)(10x-9)$

21. $(4x-9)(5x+4)$

22. $(x-5)(x^2+3)$

23. $(x+4)(5x^2-3)$

24. $(x+5)(3x-2)(3x+2)$

25. $4n(9n-7)(8n+9)$

Chapter 8

1. $x=0, 4$

2. $x=\text{-}2/3, 3/5$

3. $x=\text{-}1/4, 6/5, 4/9$

4. $x=\text{-}1/3, 2$

5. $x=2/3, 1$

6. $x=\text{-}2/5, 1/6$

7. $x=\text{-}8/15, 1/3$

8. $x=\text{-}3\pm\sqrt{14}$

9. $x=\dfrac{\text{-}7\pm\sqrt{69}}{2}$

10. $x = -5/2, 1/2$

11. $x = -8/3, 5/4$

12. $x = \dfrac{-17 \pm \sqrt{2641}}{24}$

13. $x = \dfrac{17 \pm \sqrt{103}}{6}$

14. $x = \dfrac{29 \pm \sqrt{61}}{30}$

15. $x = \pm 3$

16. $x = 0, 8$

17. $x = 3/5, 4/3$

18. $x = -7/3, 5/6$

19. $x = -25/14, 4/3$

Chapter 9

1. Axis: $x = 1$; Vertex: (1, 5); Maximum

2. Axis: $x = -2$; Vertex: (-2, -9); Minimun

3. Axis: $x = 4$; Vertex: (4, -5); Minimun

4. Axis: $x = 2.4$; Vertex: (2.4, 59.8); Maximum

5. Axis: $x = 4/3$; Vertex: (4/3, 10/3); Maximum

6. Axis: $x = 3.5$; Vertex: (3.5, 290); Maximum

7. $y = -2(x - 1)^2 - 1$

8. $y = -2(x-1)^2 + 5$

9. $y = 2(x+2)^2 - -9$

10. $y = -5(x-2.4)^2 + 59.8$

11. $y = -3(x-4/3)^2 + 10/3$

12. $y = -16(x-3.5)^2 + 290$

13. -0.581, 2.581

14. -4.121, 0.121

15. 0.834, 7.162

16. 1.058, 5.858

17. 0.279, 2.387

18. -0.757, 7.757

19. $(3x-7)(4x+9)$

20. Center: (-4, -3) $r = 5$

21. Center: (10, -6) $r = 11$

22. Center: (-2, 7/2) $r = 1/2\sqrt{190}$

23. a. $t = 4.5$ seconds

　　b. $h = 404$ feet

　　c. $t = 0.516, 8.484$ seconds

　　d. $t = 9.204$ seconds

　　e. $t = 9.525$ seconds

Chapter 10

1. (-1, 5); (1, -3)

2. (-0.2, 2.4); (1, 0)

3. (-3.536, 3.5); (3.536, 3.5)

4. (2, -9); (4, -5)

5. (-2, -18); (1, 15)

6. (-2, -3); (1, 4.5)

7. (-8, -6); (6, 8)

8. (-3, 4); (0, -5); (3, 4)

9. (-6, -8); (6, 8)

10. a. 1400

    b. 1,960,000

    c. 3,000,000

    d. 62,500

Chapter 11

1. $\dfrac{251}{384}$

2. $\dfrac{41}{360}$

3. $\dfrac{7}{15}$

4. $\dfrac{1}{4}$

5. $\dfrac{16}{11}$

6. $\dfrac{16}{17}$

7. 4

8. $\dfrac{5x^2 + x - 2}{(x-3)(x-2)(2x+1)}$

9. $\dfrac{4x^2 + 9}{(x+1)(2x-1)(2x+3)}$

10. $\dfrac{29x^2 + 23x - 10}{(x+2)(5x-2)(5x+2)}$

11. $\dfrac{72x^2 - 71x + 18}{(x+1)(3x-5)(7x-4)}$

12. $\dfrac{3x^2 - x - 5}{(x-3)(x-2)(2x+1)}$

13. $\dfrac{2x^2 - 17x + 1}{(x-3)(x+1)(2x-1)}$

14. $\dfrac{23x^2 + 9x - 14}{(x+2)(5x-2)(5x+2)}$

15. $\dfrac{-71x^2 - 77x - 40}{(x-1)(7x+5)(9x+5)}$

16. $\dfrac{3x+1}{3x+5}$

17. $\dfrac{3x-2}{3x-1}$

18. $\dfrac{2x-5}{3x-1}$

19. $\dfrac{2x-7}{3x-5}$

20. $\dfrac{x+3}{x+5}$

Chapter 12

1. 17

2. 15

3. 1

4. $x^2 + 9$

5. $5\sqrt{3}$

6. $2\sqrt{13}$

7. $2\sqrt[3]{9}$

8. $2\sqrt[5]{3}$

9. $4x^2y^3$

10. $5xy^2\sqrt{3xy}$

11. $3x^2y^4$

12. $-4xy^3\sqrt[3]{xy}$

13. $18\sqrt{3}$

14. $64\sqrt{2}$

15. $19\sqrt[3]{2}$

16. $125x^6$

17. $\dfrac{8y^6}{27x^9}$

18. $x = -14, 4$

19. $7, 15$

20. $-1\ 3/4, 3/4$

21. $-7/5, 3$

22. $x < -1$ or $x > 5$

23. $-10 \le p \le -2$

24. $-4 < v < 2/3$

25. $c \le 2$ or $c \ge 7$

26. $|\text{ticket} - 60| \leq 25$

27. $x \geq$ -4/5

28. Real numbers

29. $x \leq$ -1 or $x \geq 5$

30. $x \leq 5/2$ or $x > 6$

31. $3\sqrt{3}$

32. $18\sqrt[3]{4}$

33. $3\left(3 - \sqrt{5}\right)$

34. 17

35. 13

36. 34

37. 7

38. 15

39. 9

Chapter 13

1. $-i$

2. $-i$

3. $10i$

4. $5i\sqrt{3}$

5. -36

6. $5+11i$

7. $-5+2i$

8. $-7+3i\sqrt{3}$

9. $17i\sqrt{2}$

10. $11i\sqrt{5}$

11. $82+39i$

12. 169

13. $40+42i$

14. $\dfrac{13}{10}+\dfrac{11}{10}i$

15. $\dfrac{-8}{7}+\dfrac{6\sqrt{6}}{7}i$

16. $\dfrac{-170}{81}-\dfrac{49\sqrt{5}}{81}i$

17. Real, rational, unequal

18. Complex conjugates

19. Real, irrational, unequal

20. Real, rational, equal

21. -3/8, 1

22. $\dfrac{5}{16} \pm \dfrac{\sqrt{71}}{16} i$

23. $\dfrac{-2}{3} \pm \dfrac{\sqrt{10}}{6}$

24. 4/3

25. $\dfrac{5}{9} \pm \dfrac{\sqrt{70}}{9}$

26. $21x^2 - 5x - 6 = 0$

27. $441x^2 + 378x + 277 = 0$

28. $x^2 - 10x + 37 = 0$

29. $3x^2 + 16x + 25 = 0$

Chapter 14

1. Translate left 2 and down 1; Domain: $x \geq$ -2; Range: $y \geq$ -1

2. Translate right 3, reflect over the x-axis and stretch from the x-axis by a factor of 2, translate up 4; Domain: $x \geq 3$; Range: $y \leq 4$

3. Reflect over the y-axis and translate up 2; Domain: $x \leq 0$; Range: $y \geq 2$

4. Translate left 3, reflect over the y-axis, reflect over the x-axis and stretch from the x-axis by a factor of 4, translate down 5; Domain: $x \leq 3$; Range: $y \leq$ -5

5. Translate right 1, reflect over the x-axis, translate up 2; Domain: reals; Range: $y \leq 2$

6. Translate left 2, stretch from the x-axis by a factor of 2, translate down 3; Domain: reals; Range: $y \geq$ -3

7. Stretch from the x-axis by a factor of 1/2 and translate up 2; Domain: reals; Range: $y \geq 2$

8. Translate right 1, reflect over the y-axis, stretch from the x-axis by a factor of 3, translate down 2; Domain: reals; Range: $y \geq$ -2

9. Translate right 1 and up 2; Domain: $x \neq 1$; Range: $y \neq 2$

10. Translate left 4, stretch from the x-axis by a factor of 2, translate down 3; Domain: $x \neq$ -4; Range: $y \neq$ -3

11. Translate right 3, reflect over the x-axis and stretch from the x-axis by a factor of 1/2, translate up 2; Domain: $x \neq 3$; Range: $y \neq 2$

12. Translate left 1, reflect over the y-axis, stretch from the x-axis by a factor of 3, translate down 4; Domain: $x \neq 1$; Range: $y \neq$ -4

13. $f(x) = -(x-3)^2 + 4$

14. $f(x) = 2 |x+3| - 5$

15. $f(x) = \dfrac{2}{x+4} + 2$

16. $f(x) = 1/2 \, (x-3)^3 - 1$

Chapter 15

1. Translate left 3 units, stretch from the x-axis by a factor of 4, translate down 2 units

2. Translate right 1 unit, reflect over the x-axis and stretch from the x-axis by a factor of 2, translate up 3 units

3. $5409.88

4. 10.35%

5. 2

6. 3

7. 11.08

8. 7.349 billion people

9. $1.03705{:}10^{12}$

10. 4:30

11. 10.5 hours

Chapter 16

1. $x > -2$

2. $x < 4/3$

3. $x > 1$

4. Real numbers

5. $\log_7 (343) = 3$

6. $\log_8 (1/4) = -2/3$

7. $\log_{25} (125) = 3/2$

8. $\log_{2/3} (8/27) = 3$

9. $2^4 = 16$

10. $9^{3/2} = 27$

11. $100^{1/2} = 10$

12. $16^{-3/4} = 1/8$

13. $2\log_3(x+3) + \log_3(x-1)$

14. $3\log(x) + 2\log(y) - 8\log(z)$

15. $4\log_4(x) - \dfrac{1}{2}\log_4(y)$

16. $7/3$

17. $3/2$

18. 0

19. 4

20. 1.21

21. 12.55

22. 2.36

23. 15.28

24. 1,995,262,315.0

25. 5,011,872.3

26. 630,957,344.5

Chapter 17

1. 87, 106, 125

2. 0, -6, -12

3. 7593.75, 5695.3125, 4271.484375

4. 405, -1215, 3645

5. 25, 36, 49

6. 21, 33, 45

7. 78, 71, 64

8. 197

9. 7380

10. 1430

11. 6055

12. 12, 48, 192

13. -6, 18, -54

14. 100,000, 10,000, 1000

15. 118,098

16. 1,048,575

17. 12,150

18. 708,576

19. 60

20. $152,695.02

Index

Note: Page numbers in **bold** indicate glossary references, and numbers in *italics* indicate answers to exercises. For exercises on specific topics, reference the chapter where the topic falls in the index entry *Exercises*.

We Have
EVERYTHING®
on Anything!

With more than 19 million copies sold, the Everything® series has become one of America's favorite resources for solving problems, learning new skills, and organizing lives. Our brand is not only recognizable—it's also welcomed.

The series is a hand-in-hand partner for people who are ready to tackle new subjects—like you!

For more information on the Everything® series, please visit *www.adamsmedia.com*

The Everything® list spans a wide range of subjects, with more than 500 titles covering 25 different categories:

Business	History	Reference
Careers	Home Improvement	Religion
Children's Storybooks	Everything Kids	Self-Help
Computers	Languages	Sports & Fitness
Cooking	Music	Travel
Crafts and Hobbies	New Age	Wedding
Education/Schools	Parenting	Writing
Games and Puzzles	Personal Finance	
Health	Pets	